Collins

Key Stage 3

Geographical Enquiry

Student Book 3

David Weatherly
Nicholas Sheehan
Rebecca Kitchen

William Collins' dream of knowledge for all began with the publication of his first book in 1819. A self-educated mill worker, he not only enriched millions of lives, but also founded a flourishing publishing house. Today, staying true to this spirit, Collins books are packed with inspiration, innovation and practical expertise. They place you at the centre of a world of possibility and give you exactly what you need to explore it.

Collins. Freedom to teach

Published by Collins
An imprint of HarperCollins*Publishers* Ltd
Westerhill Road
Bishopbriggs
Glasgow G64 2QT
www.harpercollins.co.uk

Browse the complete Collins catalogue at www.collinseducation.com

First edition 2015

10 9 8 7 6 5 4 3 2 1

ISBN 978-0-00-741118-4

A catalogue record for this book is available from the British Library

Typeset, designed, edited and proofread by Palimpsest Book Production Ltd, Falkirk, Stirlingshire
Cover designs by Angela English

Printed and bound by L.E.G.O S.p.A., Italy

The mapping in this publication is generated from Collins Bartholomew digital databases.
Collins Bartholomew, the UK's leading independent geographical information supplier, can provide a digital, custom, and premium mapping service to a variety of markets.
For further information:
Tel: +44 (0)208 307 4515
e-mail: collinsbartholomew@harpercollins.co.uk

Visit our websites at: www.collins.co.uk or www.collinsbartholomew.com

If you would like to comment on any aspect of this book, please contact us at the above address or online.
e-mail: collinsmaps@harpercollins.co.uk

Contents

About this book

Space-age geography

Scheduled landing site for Philae *named* Agilkia *on comet 67P*

As we were putting the finishing touches to this third book of geographical enquiries, a remarkable thing happened. After ten years and half a billion miles of travel, the lander module *Philae* – a child of the European Space Agency – detached from its mother space probe orbiter, *Rosetta*, and touched down on comet 67P (Churyumov-Gerasimenko). Together they will complete the most detailed study of a comet ever attempted. Averaging just a few kilometres wide, with temperatures as low as –240°C, comets provide us with a glimpse of the origin of the solar system 4.5 billion years ago and might also give us clues as to how water came to be the driving force of life on Earth.

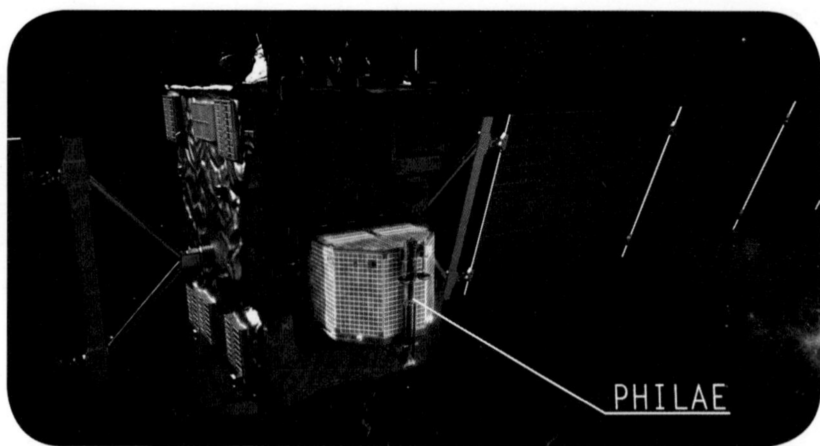

Dennis Tito

The World Tourism Organization defines tourists as people 'traveling to and staying in places outside their usual environment for not more than one consecutive year for leisure, business and other purposes'. Meet two very special tourists – Dennis Tito and Anousheh Ansari.

Anousheh Ansari

They may not be instantly recognisable but they are famous nevertheless. Tourism can be both domestic (occurring within the country a person lives) or international. Since 2002, it has also become galactic. Dennis and Anousheh were the first male and female space tourists, both enjoying a fortnight or so on the International Space Station at a cost of US$30 million each (breakfast included).

Philae*'s touchdown on the comet (artist's depiction)*

Geographers study tourism and tourists for various reasons: not only does tourism epitomise the relationship between people and their environments (which is, of course, what geography is all about), today it is also a crucial source of income for many countries. Consequently, it is vitally important to hundreds of millions of people. Worldwide, over US$1 trillion a year is generated from tourism. In 2012, China became the largest spender in international tourism in the world, spending US$102 billion and surpassing both Germany and the United States of America. Tourism is one of the top five export categories for as many as 83% of the world's countries and is a main source of foreign exchange earnings for at least 38% of all nations.

But where does space tourism leave the subject of geography which, originating from the Greek, is defined as 'earth description' or 'to describe and write about the earth'. Is geography just the study of the earth then? Does this mean that we will have to invent a new subject focusing on people – and the social, economic and environmental patterns and processes they create – in space? No, of course not. At its heart, geography is the study of the interrelationship of people with their environments, wherever they are.

One of the enquiries we have written in this book does in fact have a space-age theme and investigates how a relatively poor country with millions of people struggling with the daily grind of poverty can justify the expense of a space programme. The challenge of desertification is the focus of another investigation, which makes a thought-provoking geographical link between a drought stricken community in southern Africa and the activities of a group of people working in one of the wettest places in Western Europe. Geographers have always loved to identify different regions or areas of the world to study and the term 'Middle East' will be familiar to you. Exploring the different ways in which people have gone about defining exactly what the Middle East is provides another intriguing enquiry.

Almost every square centimetre of the land surface of the earth belongs to one country or another but what about the world's oceans? Have you ever paused to think about who owns them – they do, after all, represent nearly three-quarters of our planet. Have you ever considered where all the flotsam and jetsam that washes up on our shorelines comes from? If you haven't then now's your chance. The oceans are under threat from many sources, so you have the opportunity to investigate these and consider whether our use of the oceans is sustainable.

Endeavouring to achieve sustainable development which improves the quality of people's lives without compromising the integrity of the environment upon which we all depend is the holy grail of the twenty-first century. A southeast Asian island is the context for investigating how feasible this actually is. Finally, we invite you to explore the world of 'fog catchers' in a South American location and to evaluate the development work of those seeking to transform through irrigation some of the most arid places in the world.

It's worth reflecting that 100 years ago air travel of any kind for leisure, let alone journeys to space, was unheard of. Of course, change may not always mean progress, but one thing is for certain – thankfully, geographers such as you will always be around to make sense of it all.

David Weatherly | Nicholas Sheehan | Rebecca Kitchen

Where in the world will your enquiries take you?

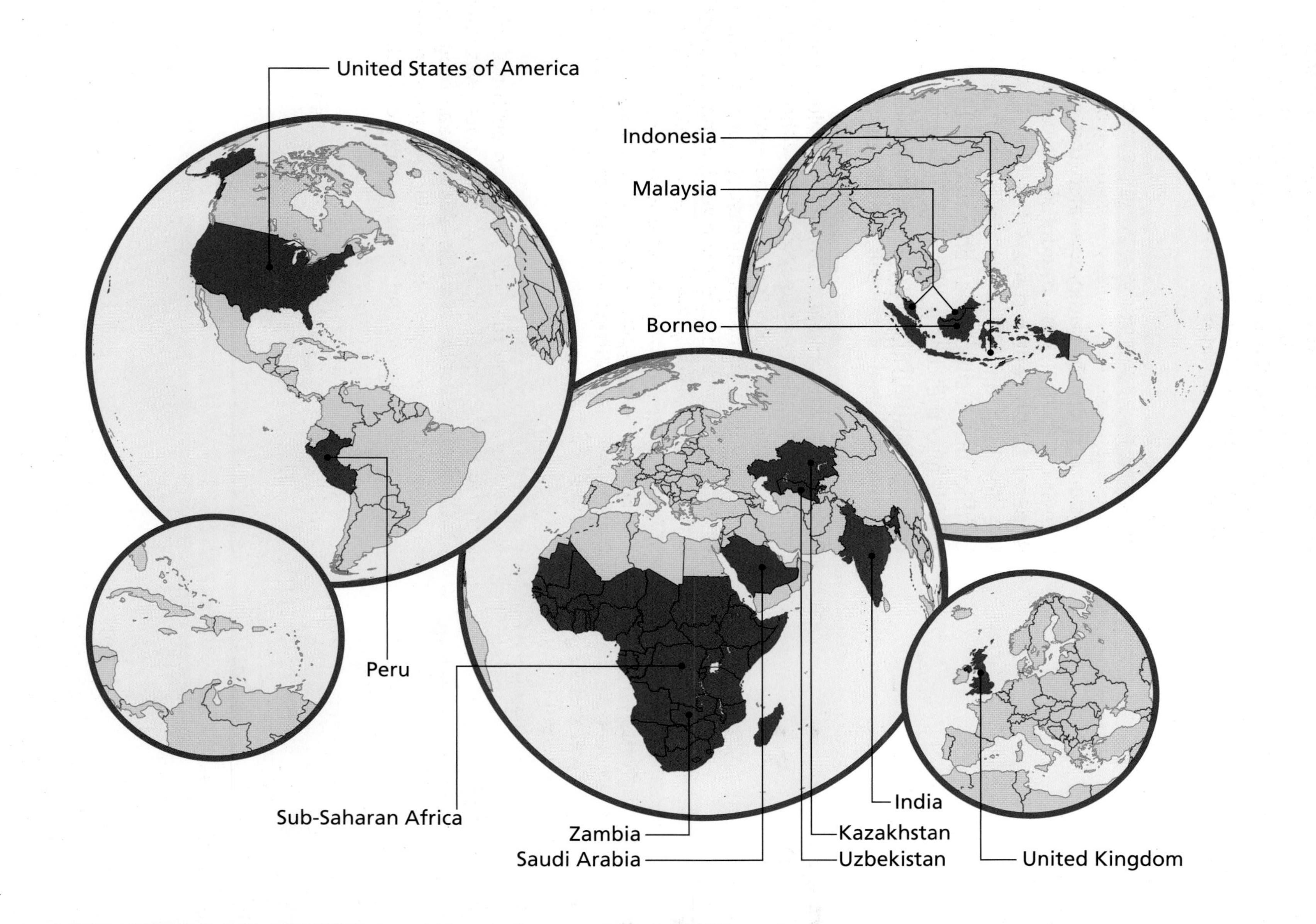

Geographical Enquiry

1 'Middle' – of what? 'East' – of where?

Is the Middle East a region?

If you turn on the radio or watch the news on television there is one **region** of the world which often appears to be centre stage – the Middle East. To some, the dominant image projected of the region is one of conflict but it should be remembered that the Middle East is also the centre of many of the world's religions and, throughout its history, has been an important cultural, political, economic and strategic centre.

As you will learn through this enquiry the Middle East is a region which is interpreted in many ways. The map below shows one of these interpretations.

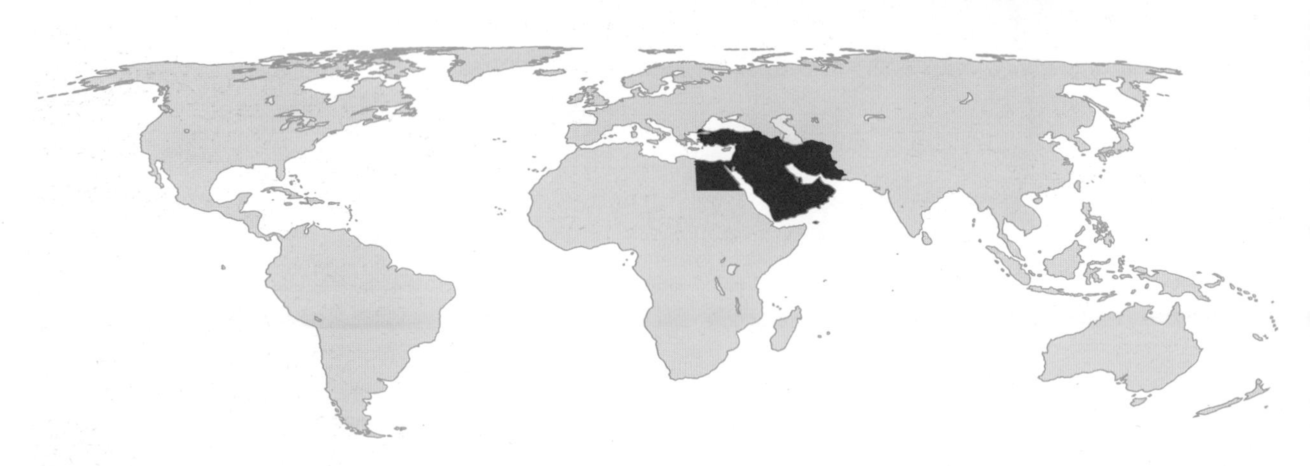

As our world becomes more globalised and **interconnected**, so events in other, perhaps fairly distant, places have more and more impact on how we live our lives. The Middle East region is likely to have a large and growing impact on our lives, yet how much do we really know about it? Why is it called the 'Middle East'? What is it like there and can we actually call it a 'region' in the first place?

1.1 How do we define the Middle East?

Alfred Thayer Mahan

The 'Middle East' is an interesting term to use to define the region, not least because the majority of people living there do not use it! The term is thought to have originated in the 1850s with the British who had **colonised** the region, but was made popular by American Alfred Thayer Mahan in 1902. The first time the United States of America used the term officially was in 1957 in a speech by President Eisenhower. However, because the term was coined by those in 'the west', i.e. Britain and America, it has been criticised for simply lumping countries together and only distinguishing them by how far they are away from Europe.

President Dwight Eisenhower

What makes the definition of the 'Middle East' even more perplexing is that it overlaps with the 'Near East' – a term which has largely gone out of use, but which was originally the maximum extent of the **Ottoman Empire**. Indeed, in 1958, the US government declared that the terms 'Near' and 'Middle' East were interchangeable.

The terms 'Near' and 'Middle' East make little sense on their own, but combining them with the term 'Far East', which describes East Asia, Southeast Asia and Siberia, makes the regional jigsaw a little clearer. However, the term 'Far East' is equally problematic as Robert Menzies the Prime Minister of Australia highlighted in 1939: *'The problems of the Pacific are different. What Great Britain calls the Far East is to us the near north.'*

President George W. Bush

A discussion of the definition of the Middle East wouldn't be complete without an acknowledgement that the terminology that we use to define it has shifted a little in recent years. The term 'Greater Middle East' was first used by President George W. Bush in 2004 to describe countries with a Muslim majority, such as Iran, Turkey, Afghanistan and Pakistan. Sometimes referred to as the 'New Middle East', the term was used at the **G8 summit** in 2004 and introduced as part of plans to change the way in which the 'West' dealt with the 'Middle East'.

Consolidating your thinking

Have a go at the learning activity on page 8 of the Teacher Book called 'Labelling places'. It will help you to understand why different places are called different things depending on who is doing the labelling. On the blank world map on page 9 of the Teacher Book, locate the places mentioned on the sheet and annotate it to explain why they are called different things. Can you find any other places that have disputed or different names? Locate and annotate these on your map. Present day Australia is a good example of a place whose name has changed over time.

- Why do you think that many people in the Middle East don't use this term to describe the region in which they live?
- Have a look at the 1644 map of Australia. What was Australia called at this time? Why do you think it was called this?

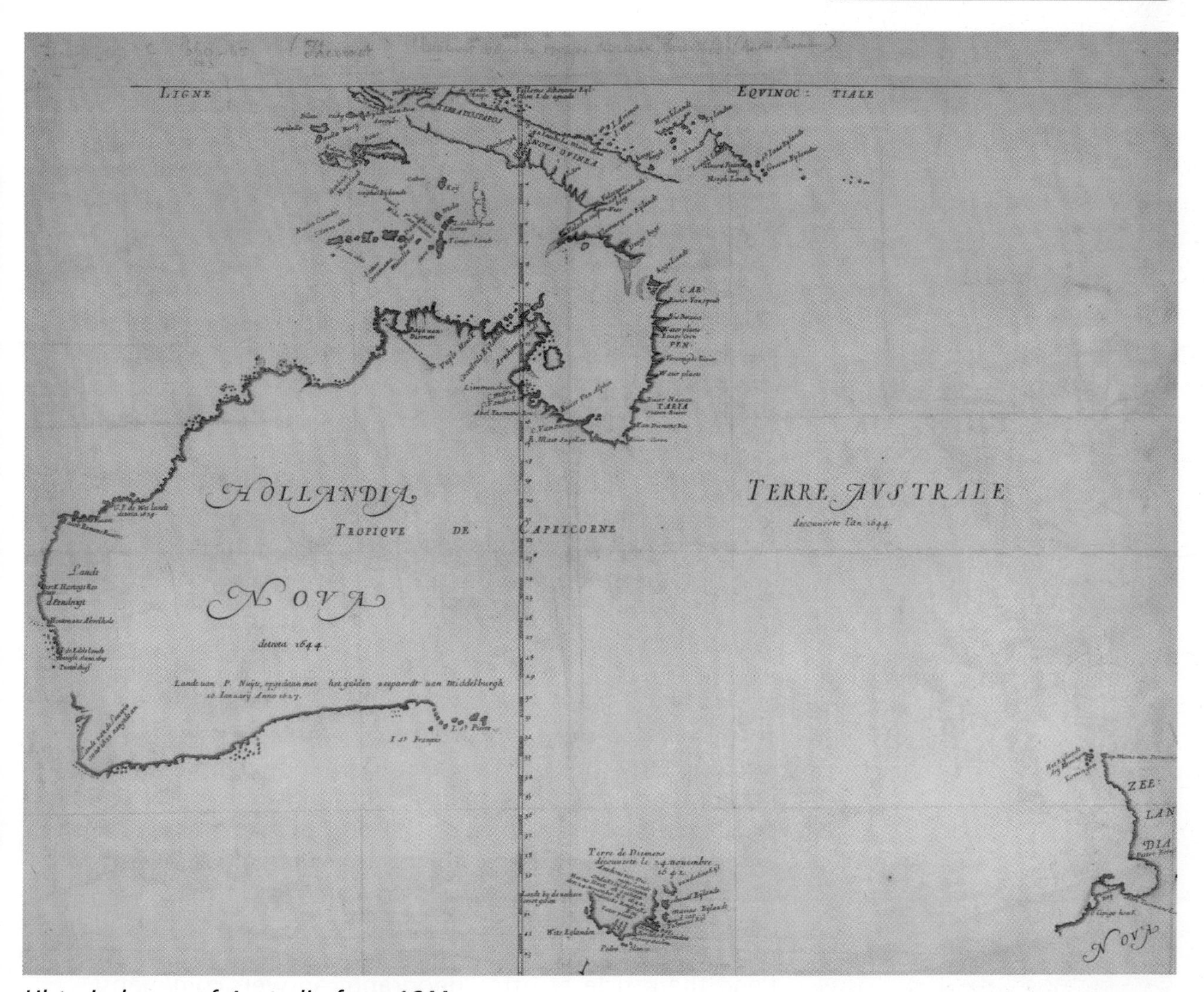

Historical map of Australia from 1644

1.2 Where is the Middle East?

If the term 'Middle East' is controversial, the location of the region is equally problematic. If you pick two different editions of the same atlas off of the bookshelf, there is no guarantee that they will define the Middle East in a consistent way (or even at all!). How is the Middle East shown in the atlas or world wall map that you use most often in geography?

Comparing the 1990 and 2005 editions of the *National Geographic Atlas of the World* highlights this. In the 1990 edition, in the section of maps entitled 'Asia', there is a map of the Middle East which shows the countries fringing the eastern Mediterranean Sea including Iraq, Syria, Jordan, the western half of Iran, the northern part of Saudi Arabia and a bit of Egypt. However, the 2005 edition splits these countries into 'Eastern Mediterranean' and 'Southwestern Asia', which includes Turkey, Iraq, Iran and Saudi Arabia plus Afghanistan and Pakistan.

The introduction to the first edition of the *National Geographic Atlas of the Middle East* starts to explain this disparity. It says:

'Fixing the Middle East with ink on paper is like reviewing a play in the middle of the second act. So volatile is the region, so unpredictable its continuing drama, that we can only set the stage and name the players. The ending – what will ultimately happen to political borders, resources, governments and peoples – is yet to be written.'

Such discrepancies in locating the Middle East really highlight the power that **cartographers** – people who make and draw maps – have over the definition, naming and classification of regions. For the purposes of this enquiry we will define the region as the area around the eastern Mediterranean Sea eastwards to Iran.

Read the article at http://geography.answers.com/maps/confusion-on-the-border-finding-countries-borders. Do you think that cartographers have a difficult job to do? What might be the consequences if they get it wrong?

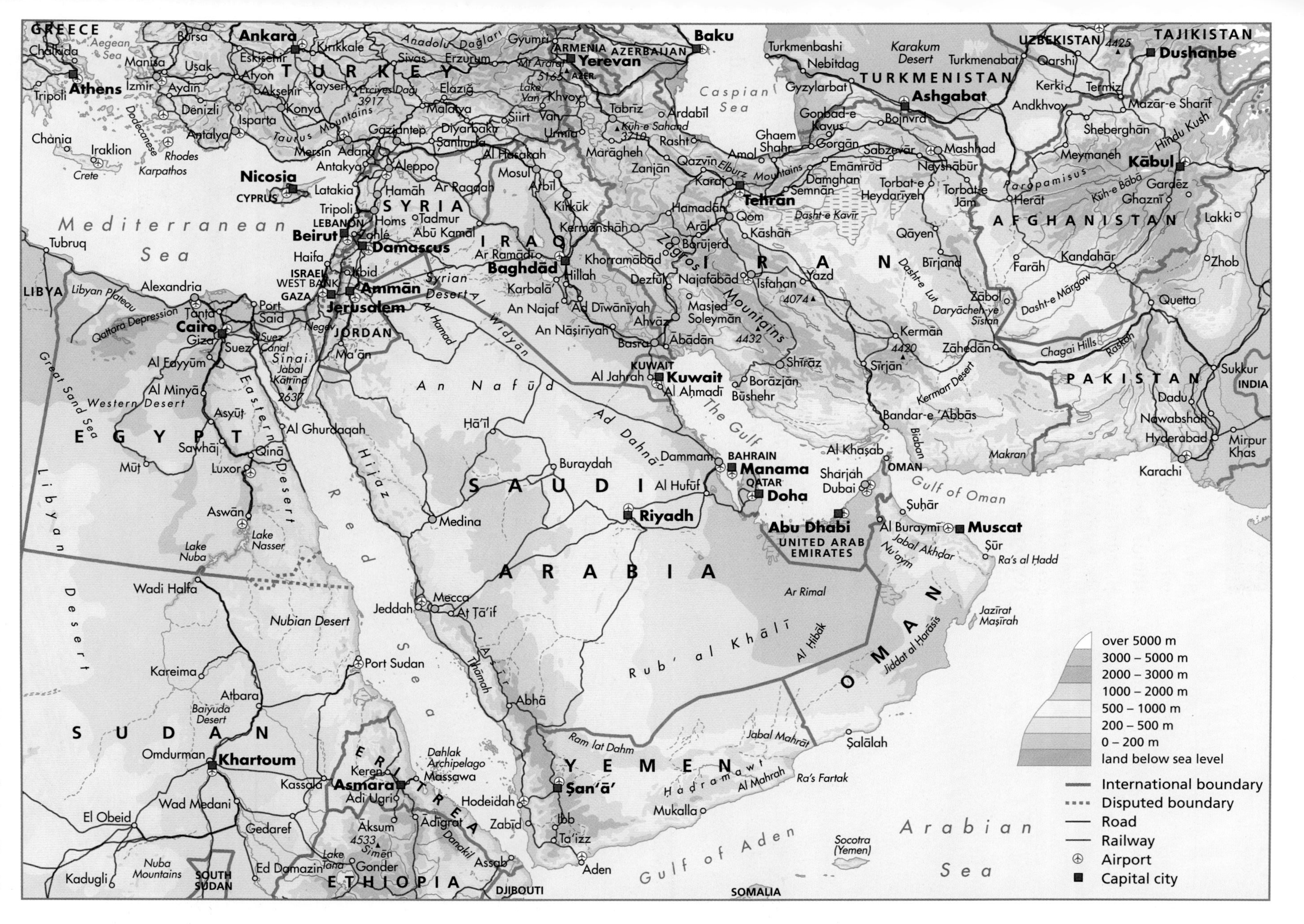
GREECE
TURKEY
SYRIA
IRAQ
IRAN
AFGHANISTAN
PAKISTAN
TURKMENISTAN
UZBEKISTAN
TAJIKISTAN
ARMENIA
AZERBAIJAN
LEBANON
ISRAEL
WEST BANK
GAZA
JORDAN
KUWAIT
BAHRAIN
QATAR
UNITED ARAB EMIRATES
OMAN
SAUDI ARABIA
YEMEN
EGYPT
LIBYA
SUDAN
SOUTH SUDAN
ERITREA
ETHIOPIA
DJIBOUTI
SOMALIA
INDIA
CYPRUS
Athens
Ankara
Nicosia
Beirut
Damascus
Ammān
Jerusalem
Baghdād
Tehrān
Kābul
Dushanbe
Ashgabat
Baku
Yerevan
Kuwait
Riyadh
Manama
Doha
Abu Dhabi
Muscat
Şan'ā'
Asmara
Khartoum
Cairo
Mediterranean Sea
Caspian Sea
Aegean Sea
Red Sea
The Gulf
Gulf of Oman
Gulf of Aden
Arabian Sea
Rub' al Khālī
An Nafūd
Ad Dahnā'
Syrian Desert
Al Hamad
Al Widyān
Libyan Desert
Western Desert
Eastern Desert
Nubian Desert
Great Sand Sea
Libyan Plateau
Qattara Depression
Karakum Desert
Dasht-e Kavīr
Dasht-e Lut
Dasht-e Mārgow
Kermān Desert
Zagros Mountains
Elburz Mountains
Taurus Mountains
Hindu Kush
Paropamisus
Chagai Hills
Hijaz
Tihāmah
'Asīr
Ḩadramawt
Makran
Sinai
Negev
Crete
Rhodes
Karpathos
Dodecanese
Dahlak Archipelago
Socotra (Yemen)
Jazīrat Maşīrah
Ra's al Ḩadd
Ra's Fartak
Alexandria
Suez
Port Said
Aswān
Luxor
Lake Nasser
Lake Nuba
Lake Tana
Mecca
Medina
Jeddah
Aden
Dubai
Sharjah
Basra
Mosul
Aleppo
Homs
Tripoli
Haifa
Izmir
Adana
Tabrīz
Işfahān
Shīrāz
Mashhad
Herāt
Kandahār
Quetta
Karachi
Hyderabad
Port Sudan
Omdurman
Wadi Halfa
over 5000 m
3000 – 5000 m
2000 – 3000 m
1000 – 2000 m
500 – 1000 m
200 – 500 m
0 – 200 m
land below sea level
International boundary
Disputed boundary
Road
Railway
Airport
Capital city

1.3 Why do geographers define regions?

So far we have referred to the Middle East as a region, but what actually is a region and why do geographers define them? Regional geography is a branch of geography that, unsurprisingly, studies the world's regions – those areas of the earth's surface with one or more similar characteristics that make it distinct. These characteristics can include the religion, economy, climate, environmental factors, politics or **topography** of the region. As well as defining regions, geographers also study and define the boundaries between these regions, known as 'transition zones'.

The **United Nations** has devised a system of dividing up the world into regions to make it easier for them to do statistical analysis. They have tried to keep the regions within the same continent and identify both large regions (which they call macro-regions) and smaller regions (which they call sub-regions). Interestingly, the Middle East is not defined as a region by the UN; rather it comes under the heading of 'Western Asia'. As we have previously seen, this is because the term is considered to be Eurocentric therefore they have chosen to use 'Western Asia' instead, although they are the only large organisation to do so. Western Asia covers exactly the same countries as the Middle East with the exception of Iran, which has been grouped with Southern Asia.

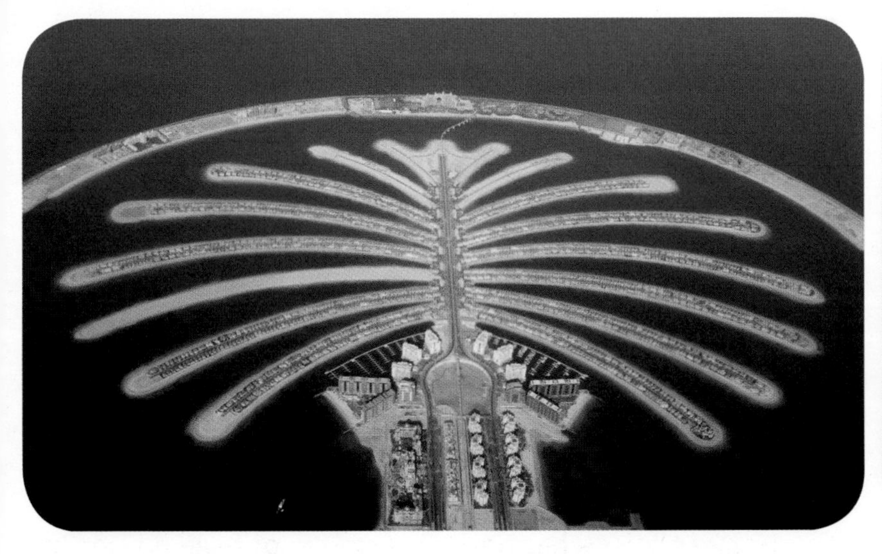

World regions as defined by the United Nations

Consolidating your thinking

- Use the learning activity in the Teacher Book on page 10 and complete the data sheet for the Middle East.
- Looking at the data, do you think that the countries of the Middle East have similar characteristics or does the data show that they are diverse?
- Look at the sheet in the Teacher Book which discusses the history of geography (p.11). Why do you think that geographers today are less interested in regions than geographers in the first half of the twentieth century?

1.4 Does the Middle East fit this definition?

If the definition of a region is 'an area that has definable characteristics which make it distinct', how far does the Middle East fit this definition? You have already looked at some data to get you to start thinking about this. Now you are going to interpret some maps to see whether or not they offer evidence of the Middle East's identity as a region.

World languages

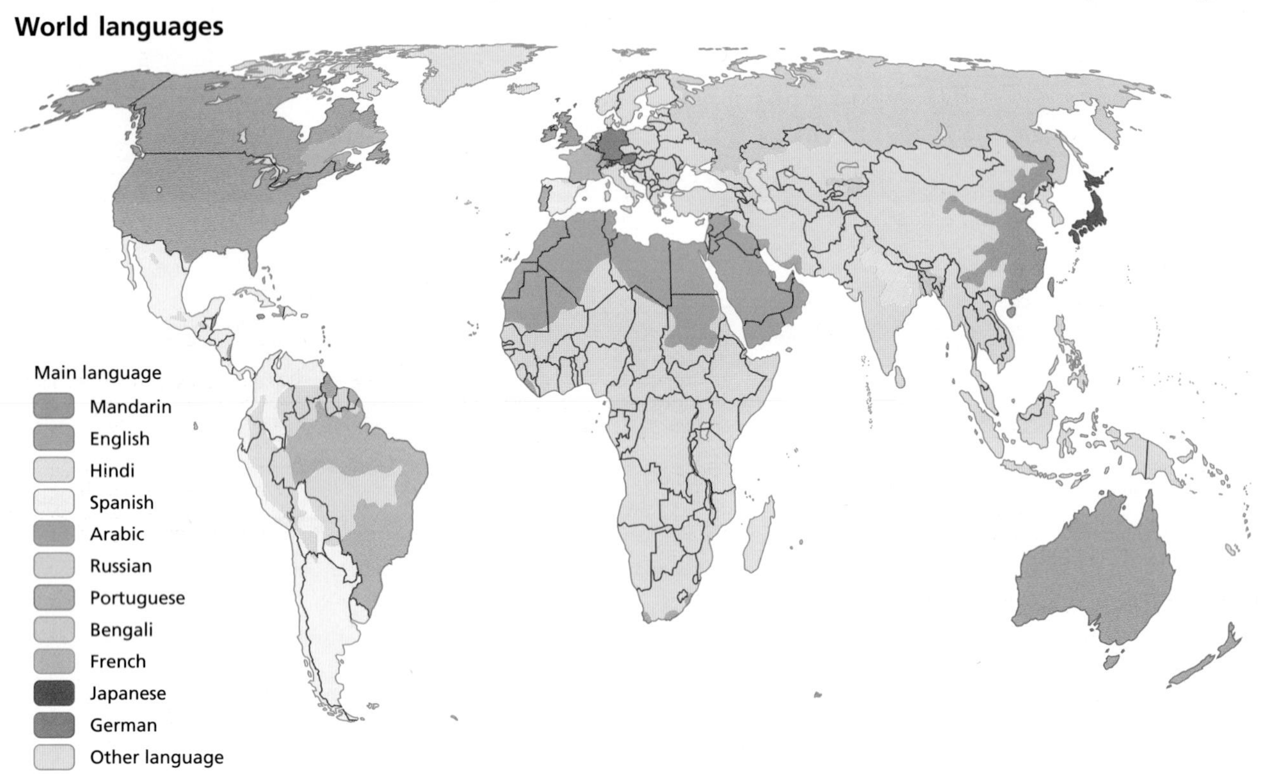

World Gross National Income (GNI)

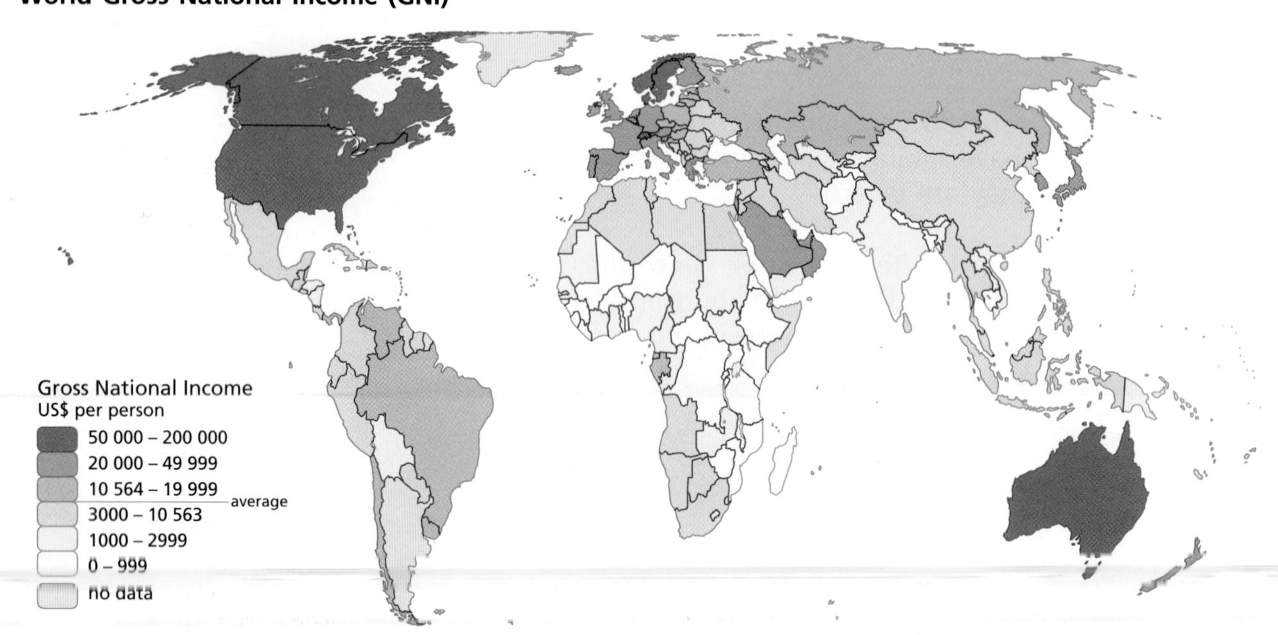

World climate

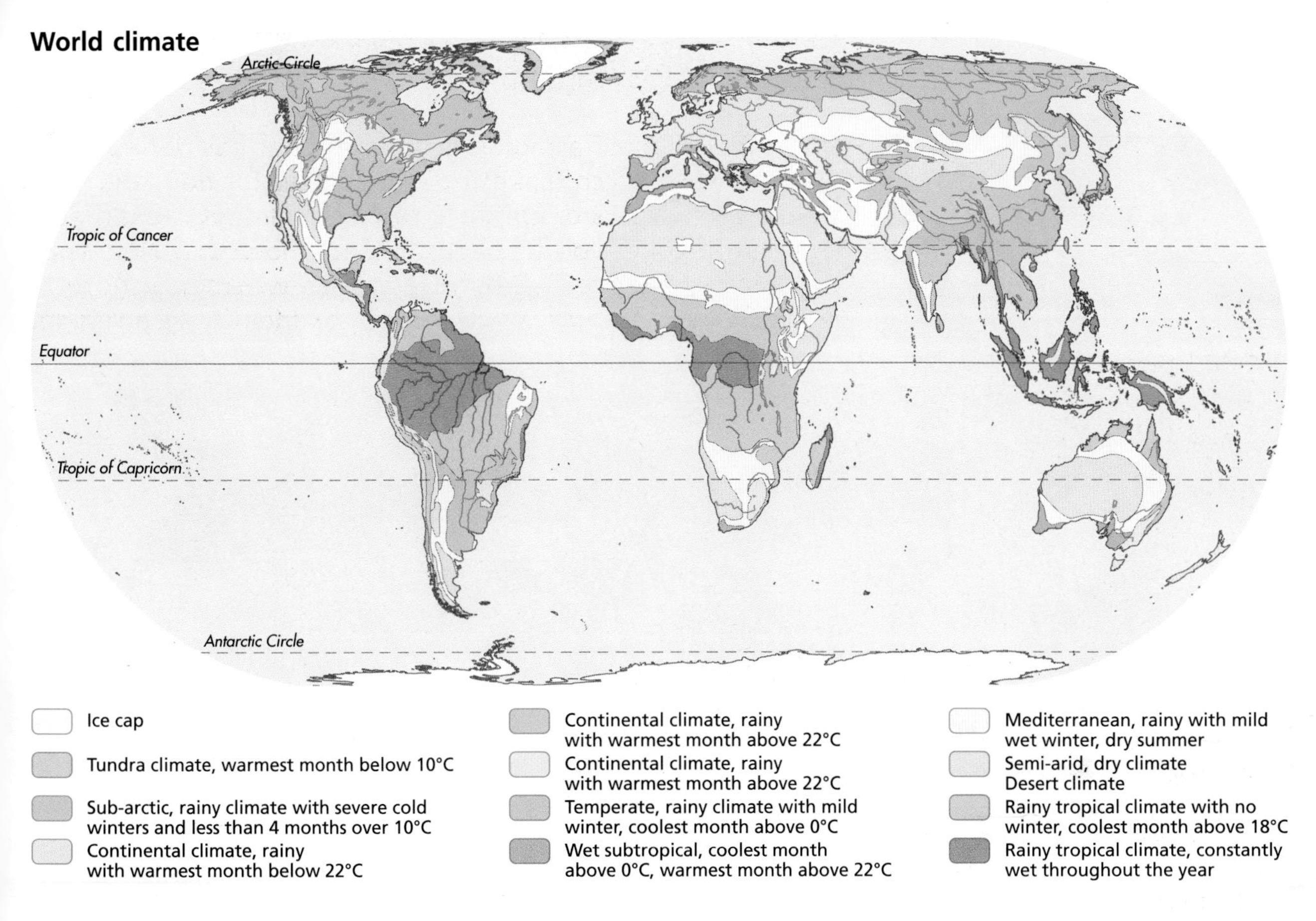

Middle East food and agriculture

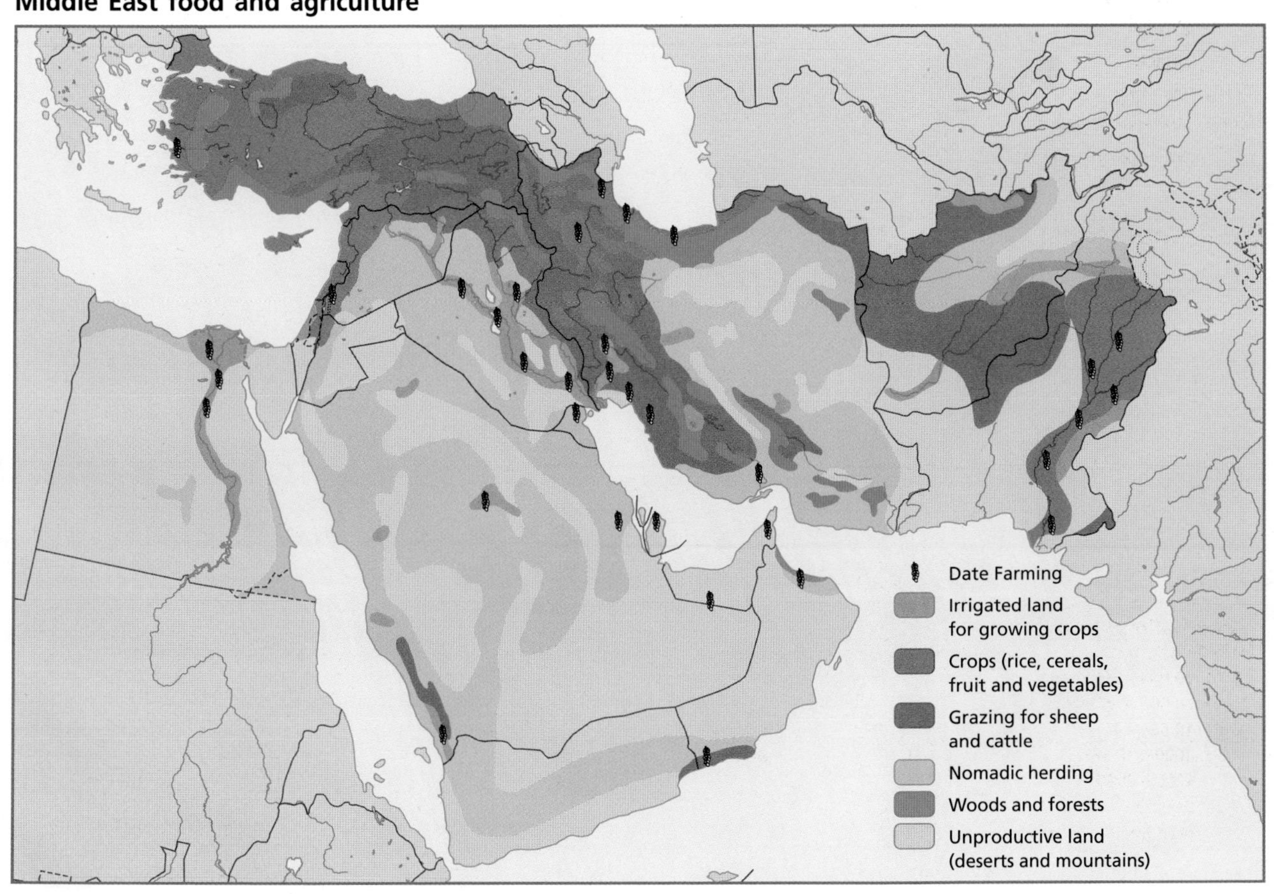

Middle East population

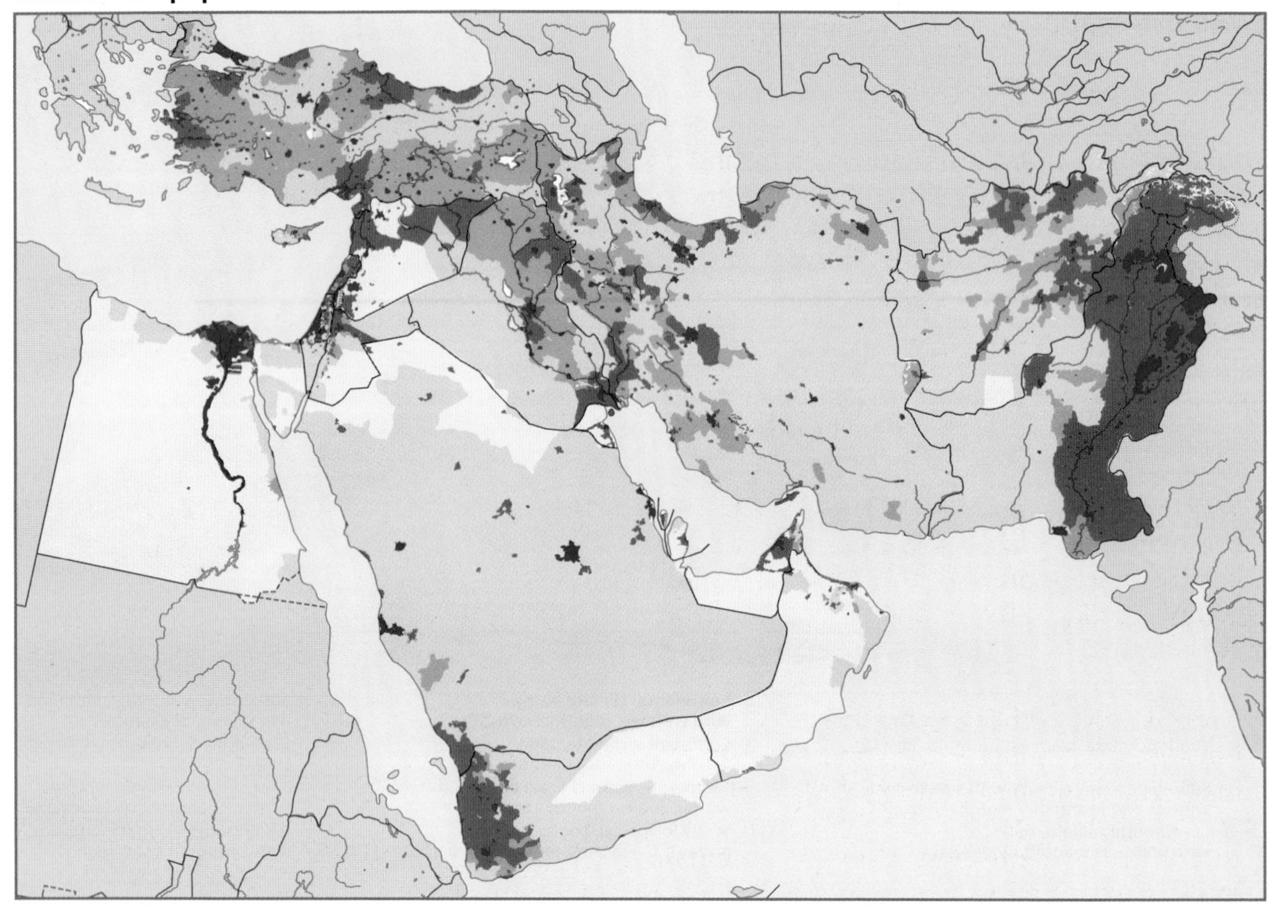

Middle East cities

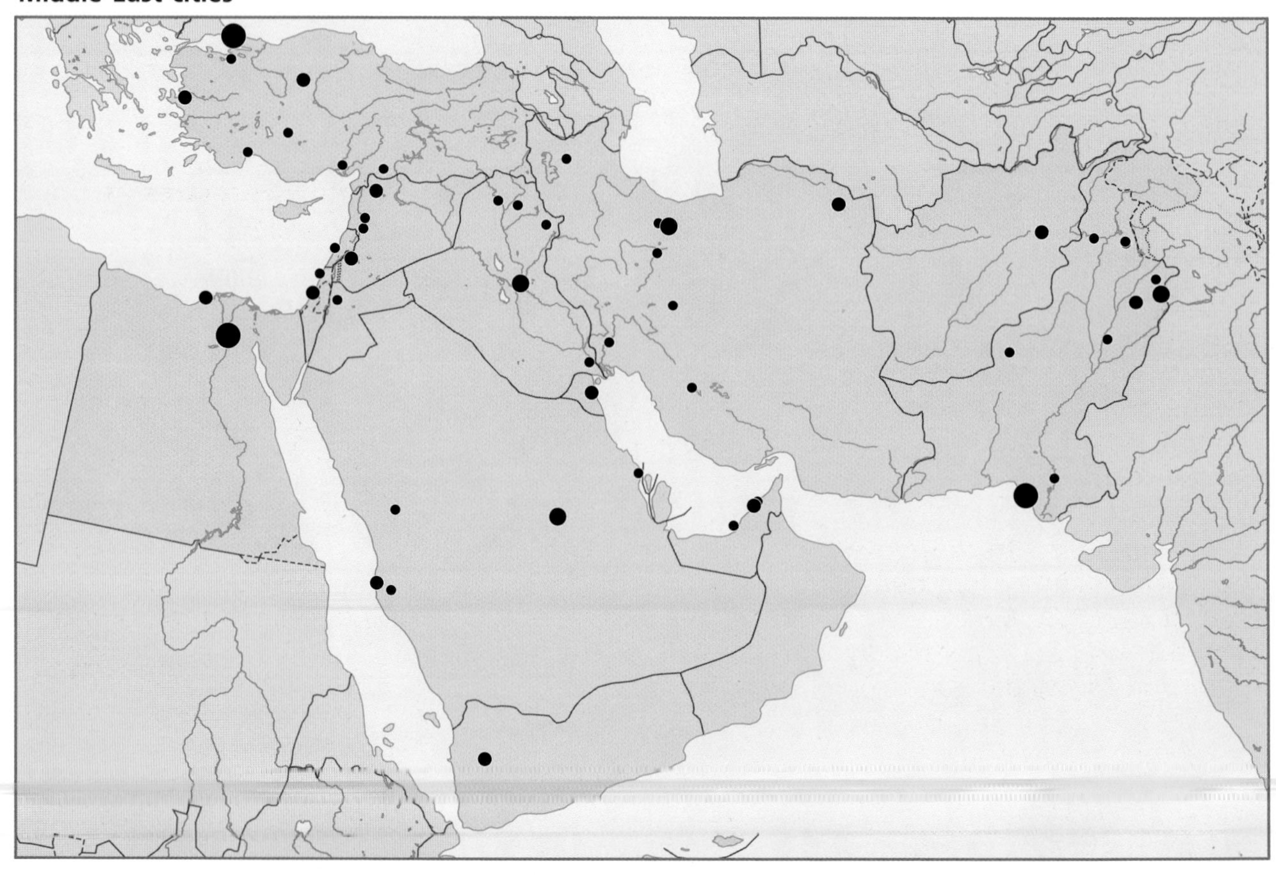

▶ Consolidating your thinking ◀

Creating a composite map

You are going to interpret some maps and create your own **composite** map of the Middle East to bring all of this information together and help you answer the key question at the beginning of this enquiry. You can use the learning activity in the Teacher Book on page 12, the information in this enquiry, plus extra reading of your own (additional sources are recommended below) to *demonstrate that you understand* whether or not the Middle East can be defined as a region.

- The first thing you need to do is select two or more of the maps which are shown on pages 16 to 18. If you choose more than two, this will make your task more complicated but it will enable you to draw more accurate conclusions.

- Once you have chosen your maps you will need to decide how you are going to combine these to make a composite map. This is probably the hardest part of the assessment as you need to get this bit right so that your map isn't too difficult to interpret. Using a combination of shading in different colours or using symbols, located bar graphs or pie charts may be a solution.

- Make sure that you create a key that is straightforward to read and put a title on your map so that it is clear what the composite map is showing.

- You then need to complete your composite map neatly. You can do this on the blank map of the Middle East in the Teacher Book on page 12.

- Once you have created your map, describe the pattern shown. Is it fairly similar across the whole area or are there large differences? Do you think that if you had chosen different maps you would have seen a similar pattern?

- Use the answers to these questions to help you write a short commentary for your map which answers the question: '*Is the Middle East a region?*' Make sure that you explain your answer fully and in detail, regularly referring to the data and your composite map.

Additional sources of information:

http://teachmideast.org/essays/27-geography/51-what-is-the-middle-east

http://www.britannica.com/EBchecked/topic/381192/Middle-East

http://www.lonelyplanet.com/middle-east

http://www.cotf.edu/earthinfo/meast/MEgeo.html

Extending your enquiry

It would be difficult to engage in an enquiry about the Middle East which didn't mention one of its most notable aspects – oil.

The Middle East is home to the world's largest accessible reserves of crude oil – a discovery which was made in 1908 in Persia (now Iran) and later in Saudi Arabia. At the time of the discovery, the world's automobile industry was expanding rapidly and oil was needed as a source of fuel. As a consequence, the kings and emirs of the region became some of the richest people in the world and the UK and USA became ever more interested in locating their oil industries in the region. Whilst oil has brought riches to the region, it has also arguably added complexity.

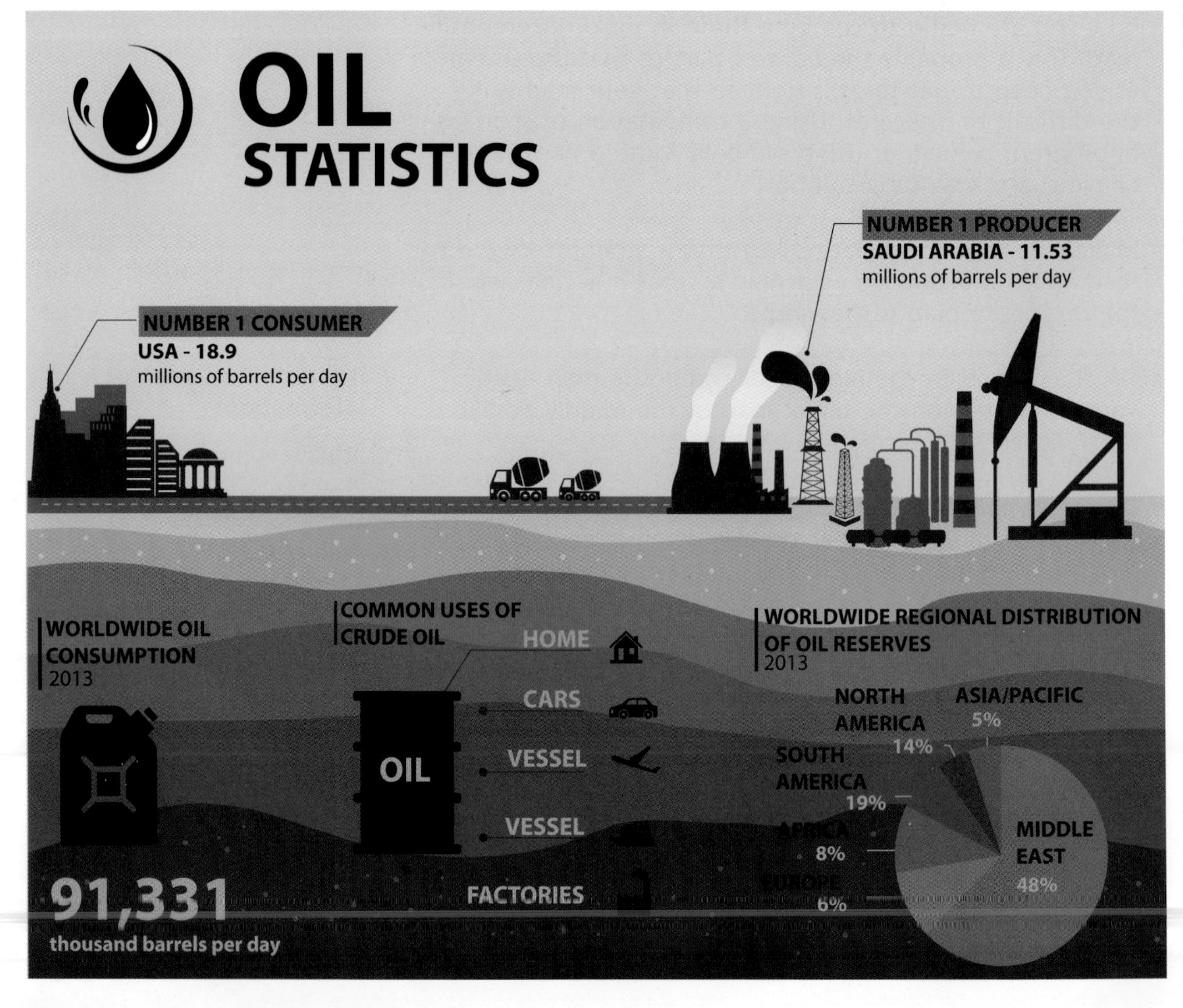

Middle East oil

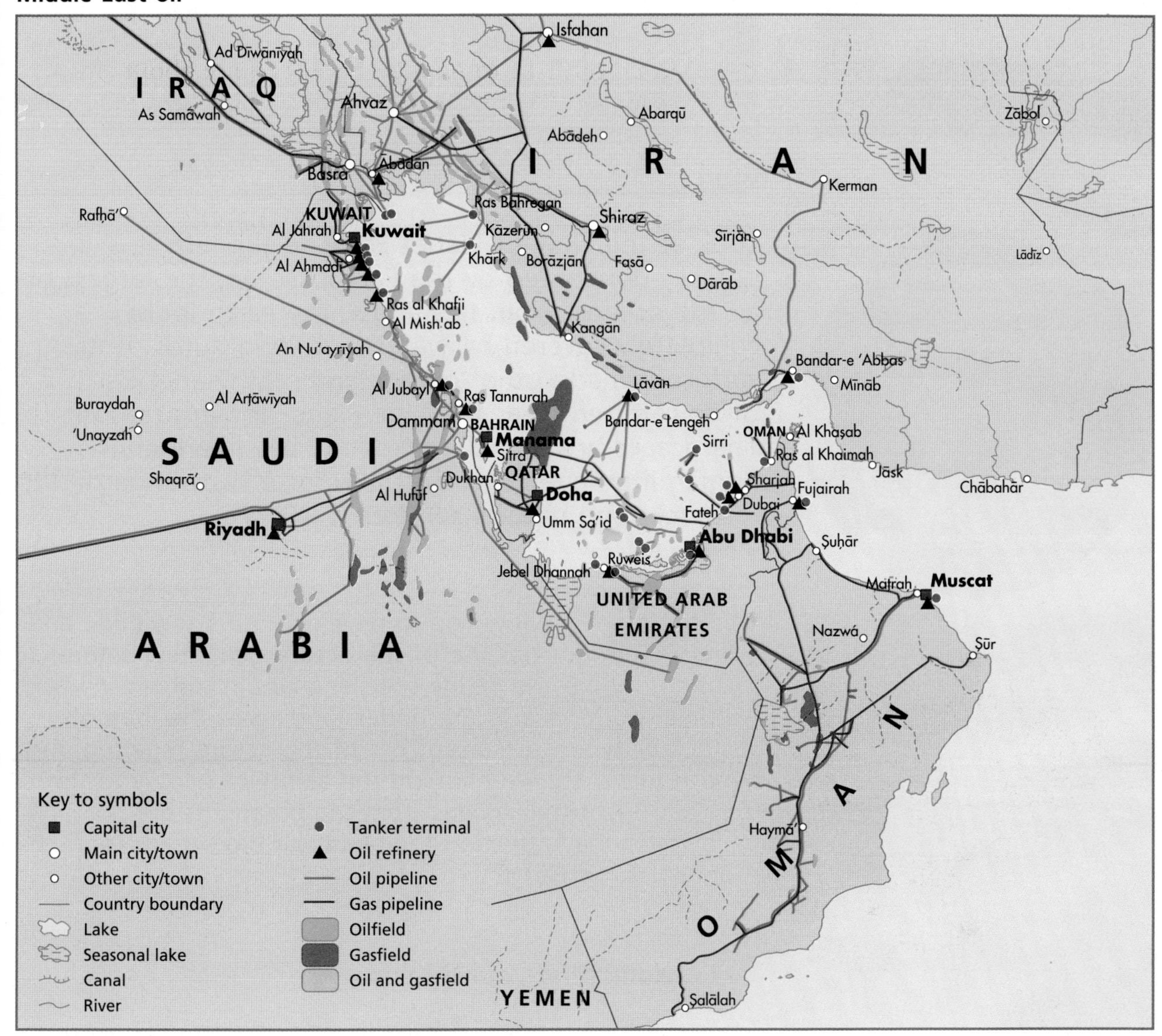

Consolidating your thinking

Read the articles in the box to the right about oil in the Middle East. Create a factfile to explain the importance of the oil industry in the region. You could include facts about the volume of oil that is **exported**, the countries that it is exported to, the impact this has had on the region and why the oil is there in the first place.

Articles on oil in the Middle East:

http://www.globalissues.org/issue/103/middle-east

http://www.bbc.co.uk/news/magazine-15037533

http://www.geoexpro.com/articles/2010/01/why-so-much-oil-in-the-middle-east

http://www.theguardian.com/environment/earth-insight/2014/mar/20/iraq-war-oil-resources-energy-peak-scarcity-economy

Geographical Enquiry

2 Issues in the oceans

Is our use of ocean resources sustainable?

Look at the view of Planet Earth to the left. It is entirely dominated by the Pacific Ocean, the largest of all the oceans. From this angle, the earth looks completely covered by water; over two-thirds of its surface is made up of oceans and seas. Planet Earth is often referred to as the Blue Planet for this reason. The oceans are a vital part of the **biosphere** and human life depends upon them for a vast range of resources and **natural services**.

This enquiry will allow you to develop your knowledge and understanding of our use of the oceans generally, before focusing in on some of the specific issues in the oceans. You will need to have a secure understanding of the idea of **sustainability**. Is our current use of the oceans meeting our needs today? Are we harming the ability of people in the future to meet their needs from the oceans? Is our behaviour today affecting the ability of the oceans to support life on Earth?

▶ Consolidating your thinking ◀

To establish your prior knowledge and understanding, you will need to complete the spider diagram on page 15 of the Teacher Book around the central question *'How do humans use and impact the oceans?'* Try to develop at least five different ideas on your own then share these with a partner.

You should then add your partner's ideas to your diagram, before asking others in the class and sharing your own ideas with them. Try to keep going until you have fifteen different uses and impacts. Once complete, try to classify your uses and impacts into different categories, such as goods, services, foodstuffs, impacts on marine animals, etc.

Oceans of the world

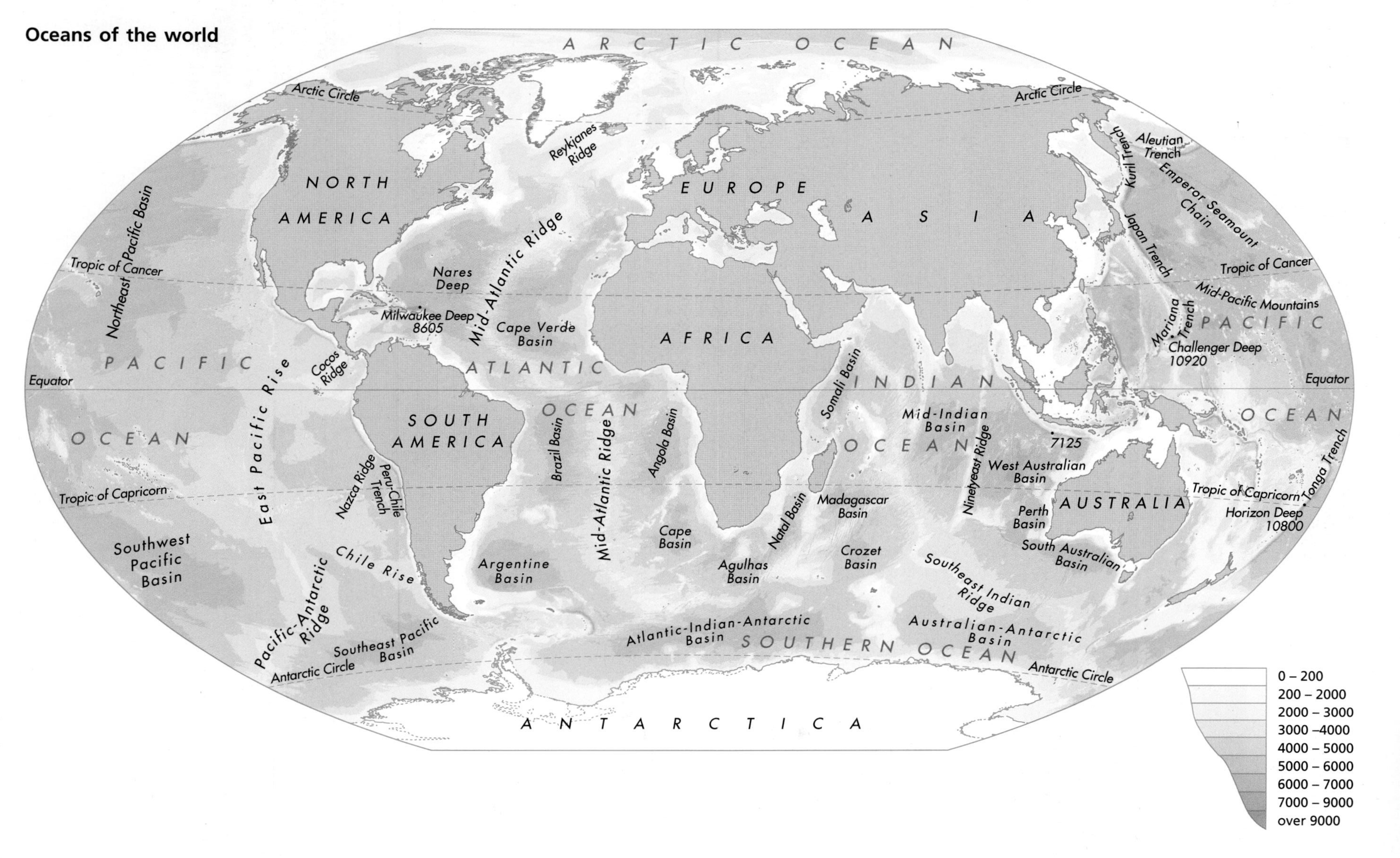

2.1 Why does Tracey keep finding Lego sea creatures on her local beach?

The oceans play a vital role in regulating the earth's climate by distributing heat from the equator and by providing us with resources like marine fisheries and oil and gas. As we continue to exploit these resources and the human population expands, we are placing increasing pressure on the **natural systems** and this is having profound impacts on the oceans – now and in the future. Some of these impacts are summarised below:

- A wide range of pollution from human activity including **nutrient pollution** from farms via rivers, plastic waste, oil, chemicals, sewage, etc.
- Rising sea levels and the impact on coasts and rates of **erosion**.
- **Overfishing** and the depletion of fish stocks.
- **Climate change** and the change in the chemistry of the ocean.
- The loss of marine species and **coral reefs**.
- Conflict over access to resources and the impact on the environment.

A clearly visible sign of our impact on the seas and oceans is **marine litter** found on the coast. Our beaches are scattered with fascinating pieces of debris. Have you ever stopped to wonder where it came from or how it got there?

Consolidating your thinking

Tracey Williams lives in Cornwall, England and for years she has been spotting interesting pieces of **flotsam and jetsam** washing up on her local beaches. In particular, she has been finding thousands of pieces of Lego – often with a nautical theme. The question is: why? People in Florida, USA have also reported finding Lego sea creatures on the shore. Could these discoveries be related?

You will need to use the images on these pages to see if you can work out why Tracey and other people are finding Lego sea creatures washing up on the beaches all over Southwest England and why people in the USA might also be likely to find some. Annotate the blank map of Southwest England on page 16 of the Teacher Book with geographical reasons that might account for the mystery.

Tracey Williams

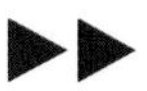

Read more about Tracey here:

http://m.bbc.com/news/magazine-28367198

https://www.facebook.com/LegoLostAtSea

http://www.bbc.com/news/magazine-28582621

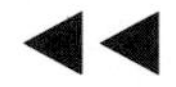

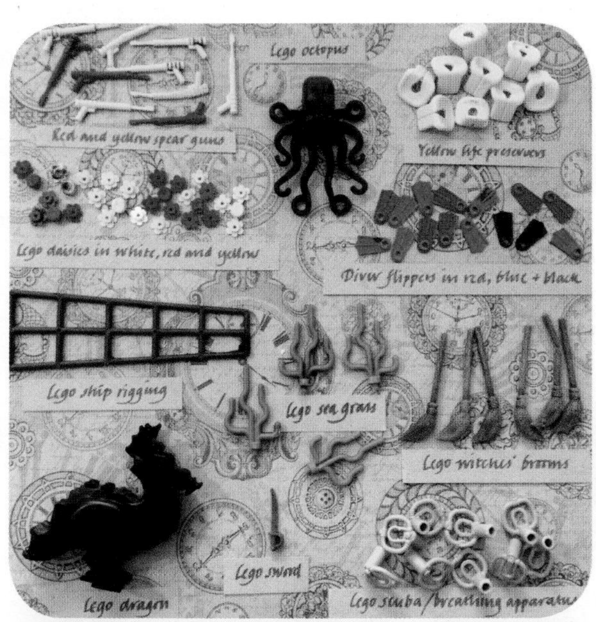

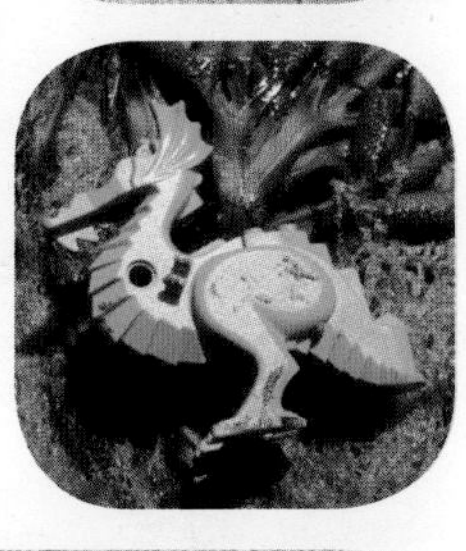

Extract from the *Southern Daily Echo*:

'The Tokio Express ***container ship*** *was bound for New York from Rotterdam, Holland in February 1997. On 13th February a freak wave hit the ship causing it to list 60 degrees one way then 40 degrees the other. The ship lost over 60 containers into the sea.'*

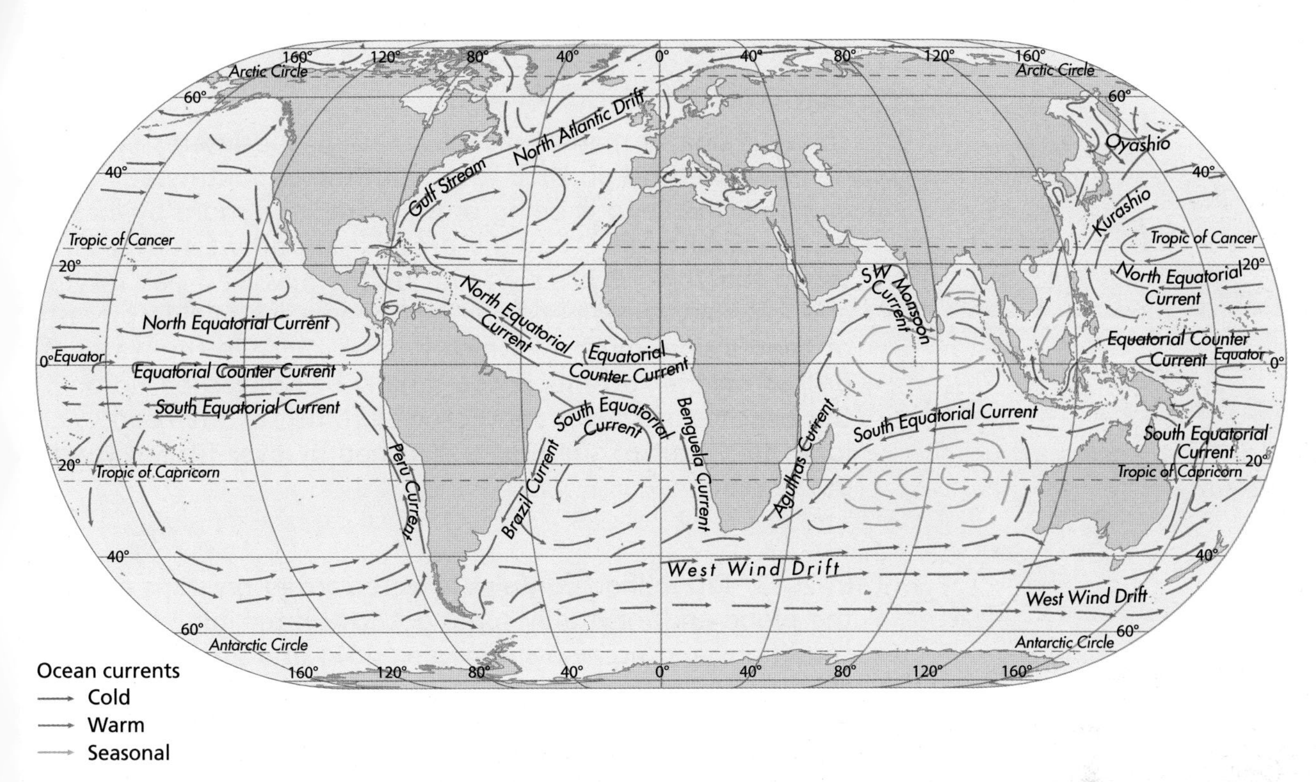
Arctic Circle
Arctic Circle
160°
120°
80°
40°
0°
40°
80°
120°
160°
60°
40°
20°
0° Equator
20°
40°
60°
North Atlantic Drift
Gulf Stream
Oyashio
Kurashio
Tropic of Cancer
Tropic of Cancer
North Equatorial Current
North Equatorial Current
North Equatorial Current
Equatorial Counter Current
Equatorial Counter Current
Equatorial Counter Current
South Equatorial Current
South Equatorial Current
South Equatorial Current
South Equatorial Current
SW Monsoon Current
Peru Current
Brazil Current
Benguela Current
Agulhas Current
Tropic of Capricorn
Tropic of Capricorn
West Wind Drift
West Wind Drift
Antarctic Circle
Antarctic Circle
Ocean currents
Cold
Warm
Seasonal

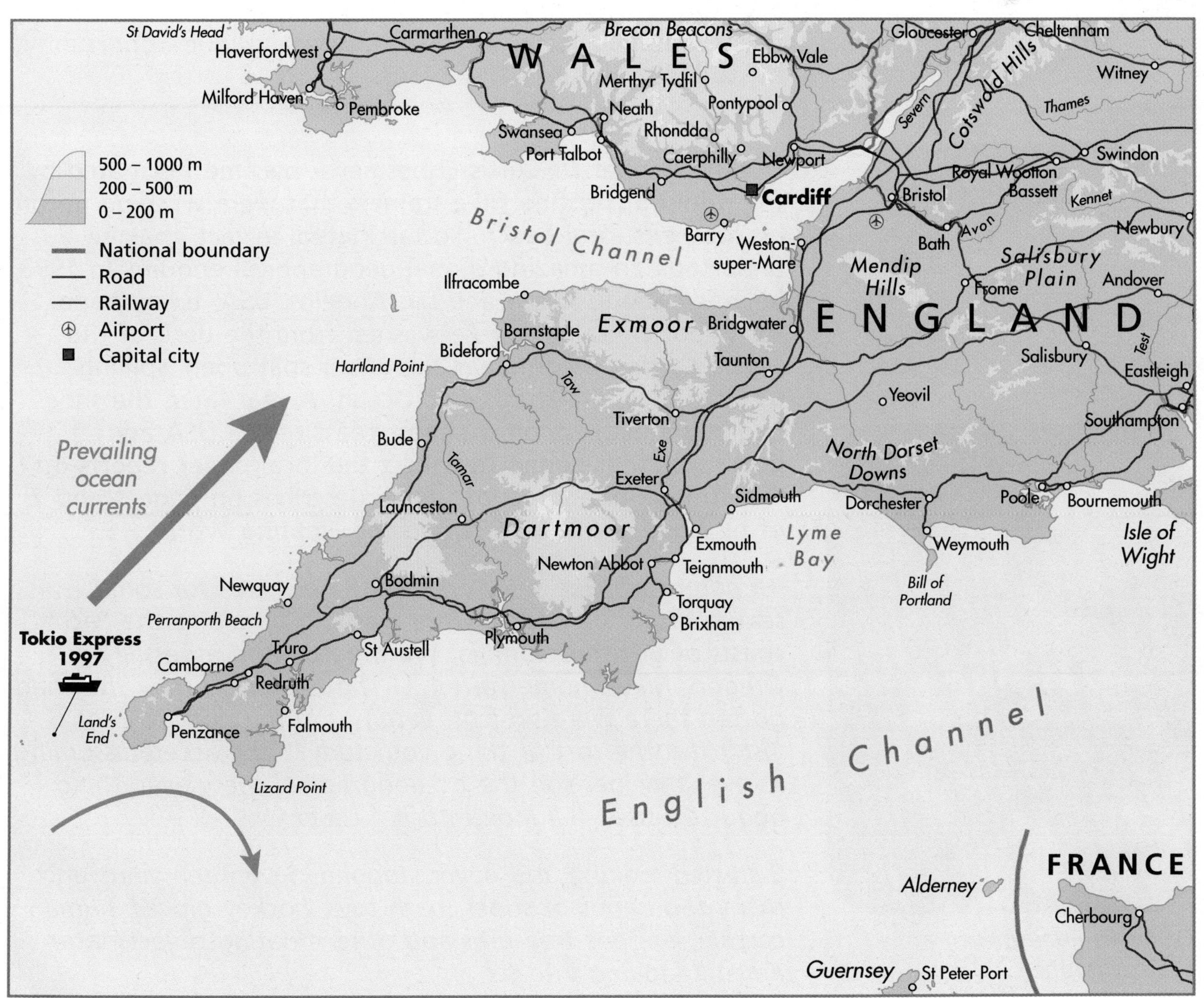
St David's Head
Haverfordwest
Milford Haven
Pembroke
Carmarthen
Brecon Beacons
WALES
Ebbw Vale
Merthyr Tydfil
Pontypool
Neath
Swansea
Port Talbot
Rhondda
Caerphilly
Newport
Bridgend
Cardiff
Barry
Gloucester
Cheltenham
Witney
Cotswold Hills
Severn
Thames
Swindon
Royal Wootton Bassett
Kennet
Bristol
Bath
Avon
Newbury
Weston-super-Mare
Mendip Hills
Salisbury Plain
Frome
Andover
500 – 1000 m
200 – 500 m
0 – 200 m
National boundary
Road
Railway
Airport
Capital city
Bristol Channel
Ilfracombe
Barnstaple
Exmoor
Bridgwater
ENGLAND
Bideford
Hartland Point
Taw
Taunton
Salisbury
Test
Eastleigh
Yeovil
Southampton
Tiverton
Bude
Exe
Tamar
Prevailing ocean currents
North Dorset Downs
Exeter
Launceston
Dartmoor
Sidmouth
Dorchester
Poole
Bournemouth
Lyme Bay
Exmouth
Newton Abbot
Teignmouth
Weymouth
Isle of Wight
Newquay
Bodmin
Torquay
Brixham
Bill of Portland
Perranporth Beach
Tokio Express 1997
Truro
St Austell
Plymouth
Camborne
Redruth
Land's End
Penzance
Falmouth
Lizard Point
English Channel
FRANCE
Alderney
Cherbourg
Guernsey
St Peter Port

Extending your enquiry

Tracey is one of many beach lovers that find interesting items washed up on their local beaches. In Southwest England, the Lego is being pushed onto the shore by local water currents driven by the prevailing southwesterly winds. When the *Tokio Express* was hit by the massive freak waves and lost sixty containers, one had nearly five million pieces of Lego inside, much of which floated. Once in the water, it drifted on the currents before finally being washed up on the beaches of Devon and Cornwall. Much of the Lego looks fresh out of the box, untouched by over fifteen years in the sea.

In 2013, 120 million containers were carried on ships across the globe and inevitably some got lost in terrible weather conditions and accidents. Between 2011 and 2013, an average of 2600 containers were lost each year. This adds up to a significant amount of material finding its way into the world's oceans. Some ocean scientists, known as **oceanographers**, have been taking a particular interest in these container losses as they provide a unique opportunity to study the oceans.

Oceanographer Dr Curtis Ebbesmeyer became fascinated by the origin of pristine Nike trainers that were washing up on his mother's local beach. So fascinated, in fact, that he undertook an amazing global geographical enquiry. In 1990, between Seoul, Korea and Los Angeles, USA, twenty-one forty-foot containers were washed from the deck of the *Hansa Carrier*, a container ship. Four split open, spilling 61,820 shoes into the Pacific Ocean. A year later, the shoes began washing up on the west coast of the USA and Curtis's mother began to collect the local paper reports on this strange phenomenon, as he describes on page 1 and 2 of his book, *Flotsametrics and the Floating World*:

'Hundreds of Nike sneakers, brand new save for some seaweed and barnacles, were washing up along the Pacific coasts of British Columbia, Washington and especially Oregon, Nike's home state [...] The details on how they had gotten there were sketchy, verging on the non existent. "Isn't this the sort of thing you study?" she asked, assuming as ever that her son the oceanographer knew everything about the sea. "I'll look into it," I said.

'I started looking and never stopped. Seventeen years and many thousands of shoes, bath toys, hockey gloves, human corpses, ancient treasures and other floating objects later, I'm still looking into it.'

Dr Curtis Ebbesmeyer, an oceanographer

Dr Ebbesmeyer began to trace the trainers by their product codes and was able to find out when and where they were lost overboard. As each shoe had a distinct serial number, it made this occurrence the largest instantaneous release of numbered objects into the ocean at one time. Knowing this and when and where they were found enabled him to build accurate maps of **ocean currents** or **gyres**. He continued his work for decades, tracking new debris and setting up a worldwide network of **beach combers**. He discovered new currents and mapped others for the first time with the **Ocean Current Surface Simulator (OSCURS)** designed by fellow oceanographer Jim Ingraham.

Use the world map from the Teacher Book (p.17) to produce your own ocean currents map. Research why the currents occur and consider what happens in the middle of each current.

A map of ocean gyres

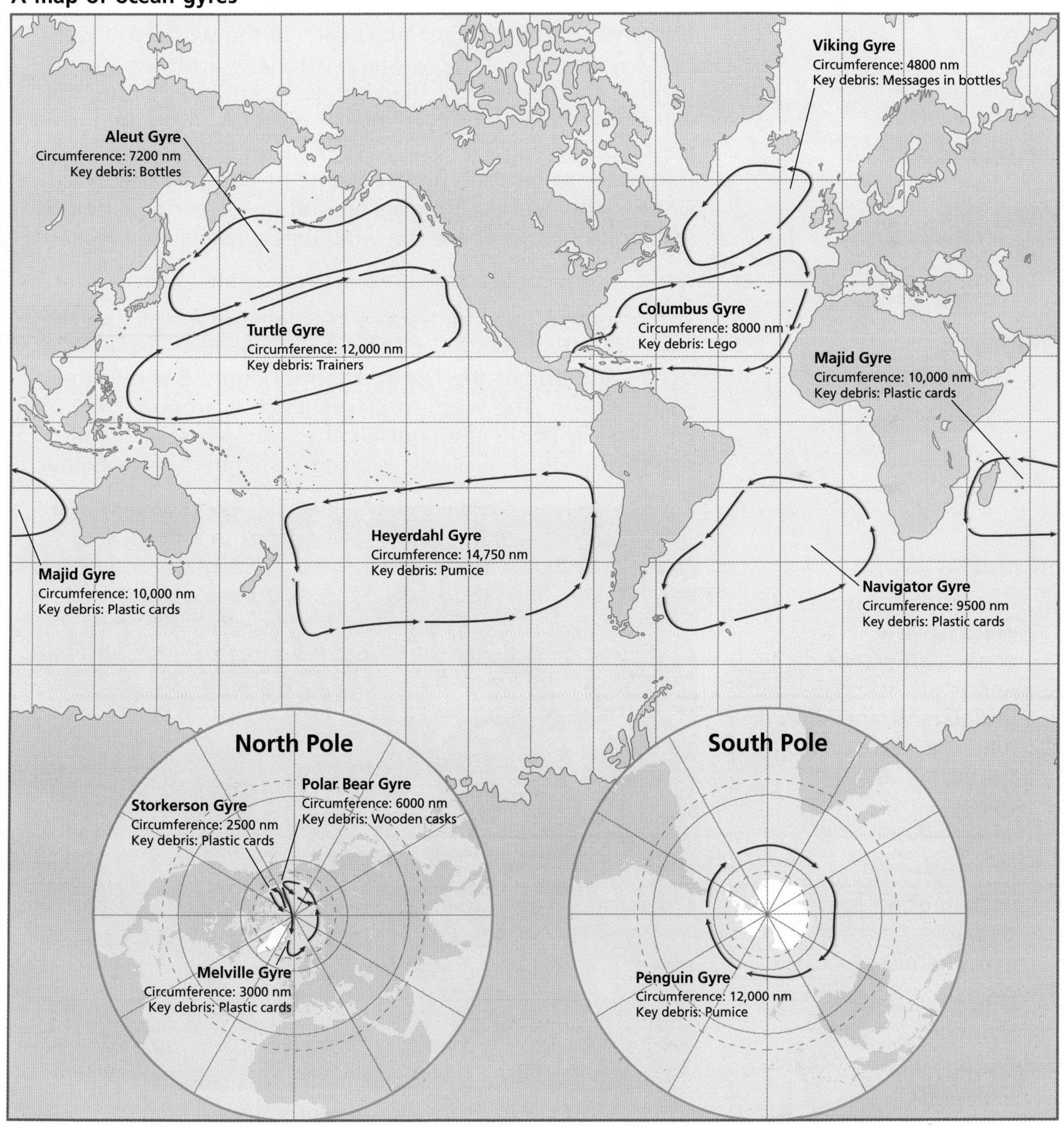

2.2 How is our use of plastic damaging the oceans?

Plastic is a resource that comes from the refining of crude oil – a finite resource. There are many different types or grades of plastic and these are often identified by a number or series of letters. In the USA, it is estimated that two million food and drink containers are used every five minutes! **Greenpeace** estimates that 100 million tonnes of plastic waste is produced each year and 10% of this ends up in the oceans.

If you walk along almost any beach in the world you will find evidence of plastic waste, particularly after big storms (Figure A). This will not **biodegrade** like natural waste, but will slowly break into ever smaller particles. A single one-litre plastic bottle will eventually break down into enough fragments to put one on every mile of beach in the world. Once in the oceans, it moves around in the currents before settling in places where the wind and currents are weaker.

The accumulations are tracked by oceanographers and they have identified two massive **'garbage patches'** in the Pacific Ocean and one in the North Atlantic. Figure B shows the Pacific Ocean. The western plastic patch is around 1,760,000 km^2 in size. This is the equivalent to an area seven times the size of the UK. The plastic is often suspended on or below

Geographical issues like plastic waste in the oceans are very important to understand and should help you develop your values and attitudes towards the environment.

Is it right that human consumption is damaging marine ecosystems?

Is our continued dumping of plastic waste into the oceans sustainable?

Figure A: *A Cornish beach after a winter storm*

the surface in a plastic soup. Up to 70% of plastic sinks, so it is also being found on the seabed. Scientists recently discovered 110 pieces of litter for every square kilometre of the North Sea – a massive 600,000 tonnes in total.

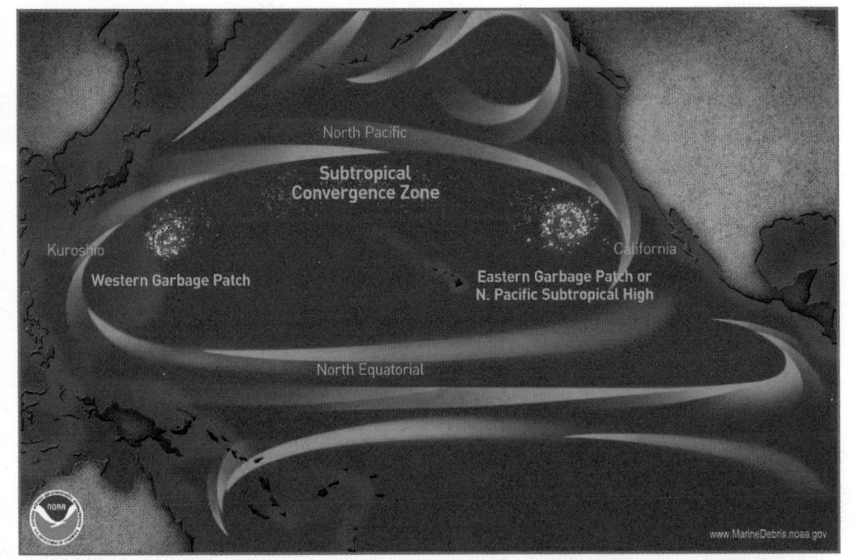

Figure B: *Location of two giant patches of waste in the Pacific Ocean*

▶ Consolidating your thinking ◀

Once in the oceans, the plastic is having hugely harmful effects on the marine environment. Figure C shows the carcass of an albatross. It has eaten a huge amount of floating plastic and died. It is just one of an estimated one million seabirds and 100,000 marine mammals killed each year by eating plastic. Huge numbers of fish are also ingesting plastic waste and we are eating the fish, so plastic waste is entering the human food chain.

Figure C: *Carcass of an albatross containing plastic waste*

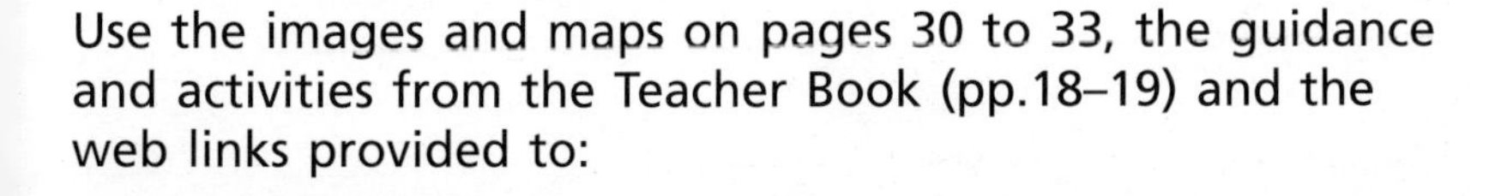

Use the images and maps on pages 30 to 33, the guidance and activities from the Teacher Book (pp.18–19) and the web links provided to:

- Research the issue of plastic waste in the oceans in detail by watching the videos and taking notes.
- Produce a simple booklet that explains the issue, its causes and consequences, aimed at 6 to 8-year-old students.
- Include pictures, maps and statistics.
- Think carefully about your use of language.
- Ensure that you include any possible solutions.

Web links to help with your research:

http://www.algalita.org/research/gis-mapping/

http://www.algalita.org/videos/videos/

http://marinedebris.noaa.gov/

http://theplastiki.com/

http://junkraft.org/

http://5gyres.org/

Accumulated plastic waste on Midway Atoll

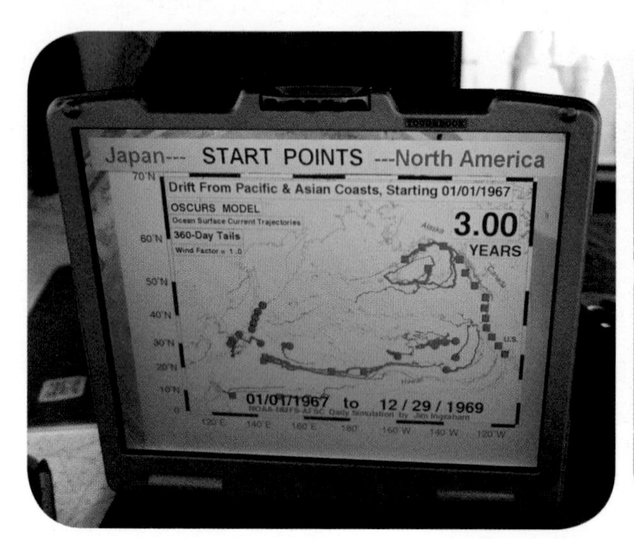

These two images are from an expedition to the North Pacific garbage patch to collect samples

A boat made from recycled plastic bottles set out on Pacific expeditions to collect data and educate people

The Great Pacific Garbage Patch

The area

The patch is around 1,760,000 square kilometres. It is around seven times the size of the United Kingdom.

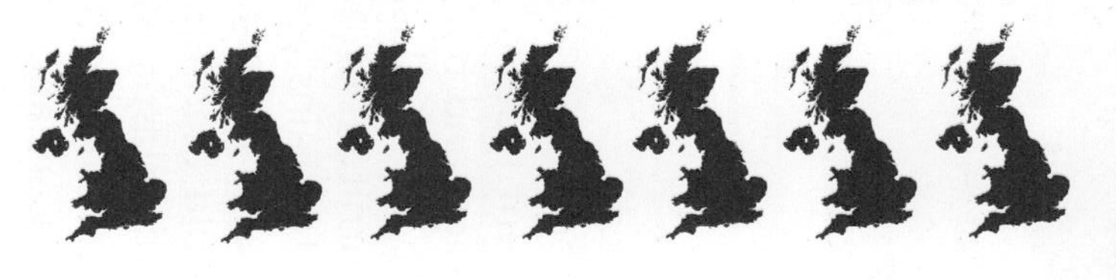

How is it formed?

Currents in the Pacific Ocean create a circular effect that pulls debris from North America, Asia and the Hawaiian Islands. They then push the debris into a floating pile containing 100 million tons of rubbish.

Where does all the rubbish come from?

20%
Ships and ocean sources like nets or fishing equipment, many containers fall into the sea after severe storms

80%
Brought by sewer systems and rivers to the sea

Photodegradation

400 years
Six-pack plastic ring

450 years
Plastic bottle

FACT

Only 6.5% of plastic is recycled every year in the USA

Plastic problems

- Plastic fouls beaches worldwide and puts off tourists
- Some plastics entangle and kill marine animals
- Washed up plastic destroys the habitats of coastal animals
- Plastic gets inside ships' propellers and keels, making ship maintenance more expensive
- Plastics are ideal vessels and enable invasive species to travel to further regions
- Most plastic never biodegrades, it takes hundreds of years to break down into smaller pieces, which are still plastic

2.3 What can you do to help?

Captain Charles Moore founded the Algalita Marine Research and Education Organisation after a voyage through the Pacific where he was horrified by the quantity of small plastic waste that he sailed through each day. Many of the web links on pages 31 and 37 are from institutes, organisations and expeditions that are trying to raise awareness of this issue in the hope of educating people and making a change in our behaviour. Individuals like Tracey Williams, Dr Ebbesmeyer and Captain Charles Moore are making a difference, backed by teams of volunteers, oceanographers and research scientists.

We can all make a difference by avoiding plastic in the things we buy and by disposing of our waste responsibly.

The Ocean Cleanup foundation (http://www.theoceancleanup.com/) are pioneering a method to extract plastic waste from the oceans using floating booms and towers. You can read more about their plans on pages 20 and 21 of the Teacher Book. We still need to tackle the problem at the source by preventing it from getting into the sea or, better still, reducing the volume of plastic waste we are producing in the first place. Why not try some of these ideas:

- If you live by the sea, you should work with friends, parents and teachers to organise a beach clean-up of your local beach – you could use Twitter or Instagram to build up support (see http://beachclean.net/).
- Use a bag for life and a reusable water bottle and ensure you recycle your plastic at home and school. If there are no facilities at school then start a campaign.
- Return unrecyclable plastic packaging to the shop it came from, the manufacturer or send it to your local politician explaining why you are doing this.

Captain Charles Moore

Extending your enquiry

Use the extra resources provided in the Teacher Book (p.22) to learn more about the different types of plastic that are commonly used. In groups, design and build a useful model that reuses as many different types of plastic as you can find. It could be a toy, some furniture or even kitchen or office products. You will need to brainstorm ideas, agree on the best idea, sketch an annotated design and discuss construction methods. You should then build the product or a scale model using the available plastic. You will need to present your model to your class, highlighting why reusing and recycling the plastic is necessary and linking it to the oceans of plastic.

2.4 Where have all the fish gone?

Plastic waste entering the oceans and accumulating into giant patches of garbage is a significant threat to the health of the oceans, to marine ecosystems and potentially to human health and wellbeing. However, the oceans are also under threat from other types of human activity.

The 2014 **World Wide Fund** for Nature (WWF) Living Planet Report identified a 39% decline in marine species (birds, mammals, reptiles and fish) between 1970 and 2010. The trends vary, but the tropical Pacific and Atlantic oceans and the Southern Ocean have seen the sharpest declines. The report states that in these regions, *'seabirds, sharks and many fish populations have seen declines as a result of **over fishing** and **bycatch**'*.

Humans have always used the oceans as a source of food. They are a valuable resource responsible for feeding billions of people each day. However a lack of management, increasing technology and the use of bigger and bigger boats has led to dangerously high levels of overfishing. In some areas, fish we love to eat are in danger of disappearing forever.

As demand for fish has increased, the profits to be made from fishing have risen. Big companies can afford to build ever bigger ships capable of taking even more fish out of the sea with each haul. Figure A shows one of the world's biggest **pelagic trawlers**. These **high-tech** vessels use radar and spotter helicopters to identify shoals of fish. Their massive nets – capable of containing a fleet of twelve jumbo jets – clear everything caught in them. This boat catches up to 8000 tonnes of tuna in each haul.

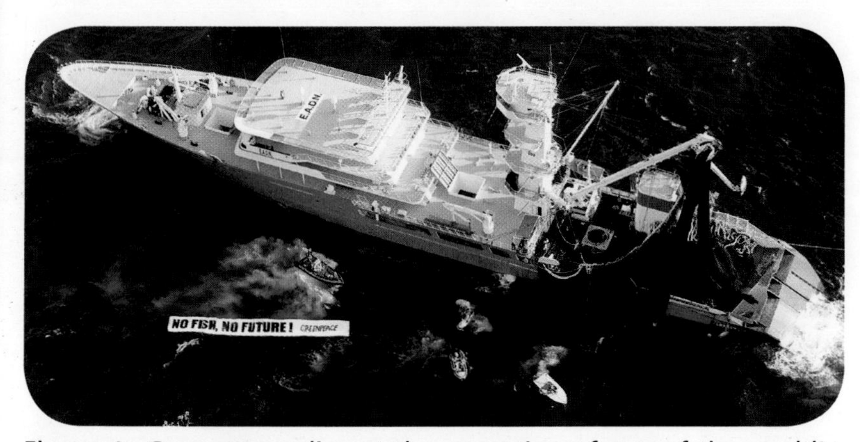

Figure A: *Greenpeace disrupt the operation of one of the world's largest fishing boats*

In the Pacific, there is huge concern over the arrival of super trawlers and floating **fish factories** such as the *Lafayette*, a Russian registered and Hong Kong owned ship that takes fishing at sea to a truly industrial scale. The ship doesn't actually catch fish, but acts as a mother ship to a fleet of super trawlers. It pumps the vast catches of these trawlers, sorts, processes and freezes the fish on board. It even has twelve forklift trucks on board to move the frozen fish around, ready to be shipped off to other ships that transport the fish to land. The *Lafayette* rarely has to dock so it can maximise its time processing fish.

Modern technology simply leaves the fish nowhere to hide and, as a consequence, some species are at risk of extinction due to overfishing. Greenpeace and the Marine Conservation Society (MCS) have identified a large number of popular species that are seriously threatened by:

- Catching of vulnerable species.
- Destructive fishing methods.
- Overfishing.
- Unselective fishing methods that create wasteful bycatch.
- Illegal or **pirate fishing**.

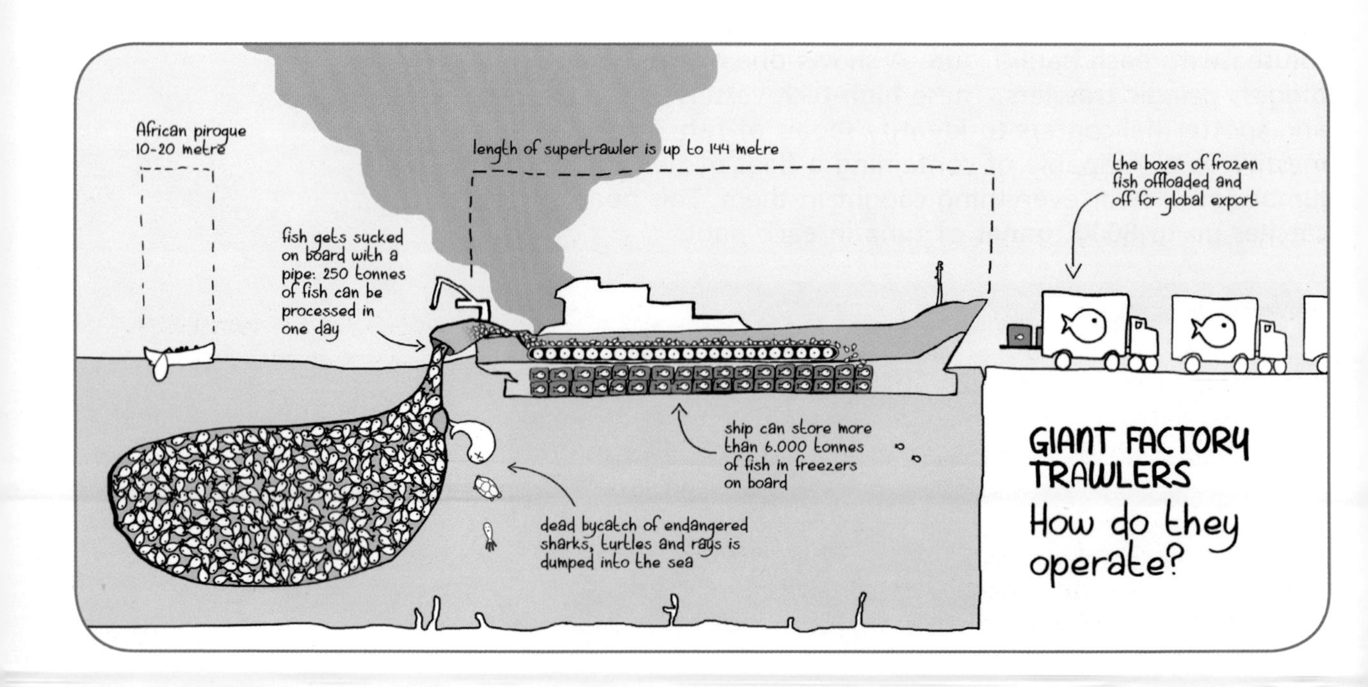

Species under threat include monkfish, Atlantic cod and salmon, haddock and tuna. There are several different species of tuna found in different oceans. The five main species are skipjack, bluefin, big eye, albacore and yellowfin. All of these are under threat in at least one location around the world. For example, skipjack tuna should not be eaten if it is from the western Atlantic, it is at risk in the Indian Ocean, and is best if caught by hand on a pole and line in the central Pacific.

When choosing fish to buy, you should consider whether *you* want to continue eating this in the future or whether you want *future generations* to enjoy what we have today. In many areas the current situation is unsustainable.

There are other methods of fishing and campaigns to raise awareness of these issues. In Figure A on page 35, Greenpeace are seen disrupting the work of the trawler. Many retailers are putting information on the labels of products and the location should always be checked. If in doubt, ask, as consumers that demand sustainably sourced fish will bring about change.

Consolidating your thinking

You will need to use the resources from the pages of this Enquiry and the Teacher Book to complete a research task. In groups, research one of the main threats to different marine species and the range of possible sustainable solutions. Plan and produce a presentation to answer the following questions:

- Where is the species found?
- Where is it under threat?
- Why is it under threat?
- What are the consequences of this?
- Are there any sustainable solutions to the problem?

Use a maximum of six slides and sixty words per slide. Allocate roles in the group – speech writer, image researcher and ICT expert, for example.

Web links to help with your research:

http://www.greenpeace.org/international/en/campaigns/oceans/fit-for-the-future/overfishing/

http://www.fishonline.org/information

3 Going boldly

Is India's space programme justified?

In November 2013, and at a cost of US$118 million, India launched an unmanned lunar orbiter on a nine month trip to Mars called *Mangalyaan* or Mars vehicle. It was the latest event in the country's rapidly growing space programme, which currently costs the government US$1 billion a year. As the space craft rose from the launch pad, India's space programme became only the fourth in the world – after Russia, the United States and Europe – to venture to another planet.

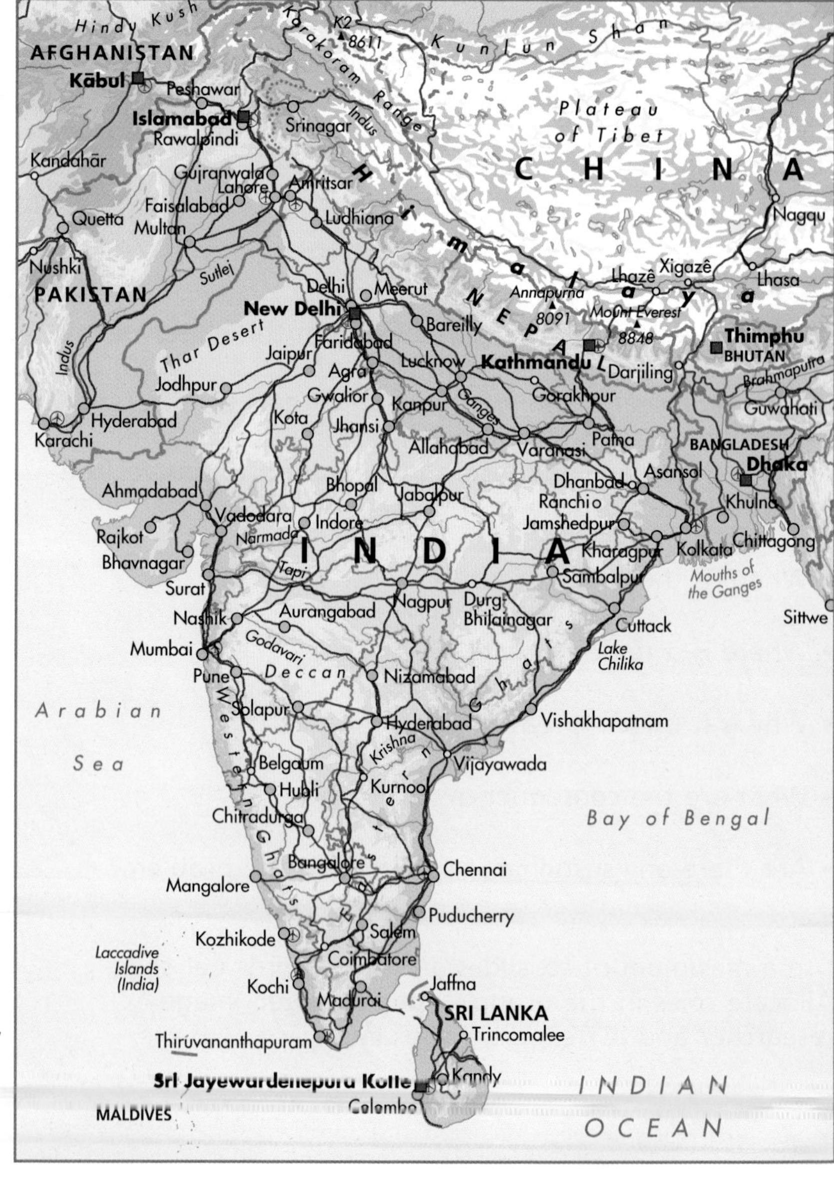

over 5000 m
3000 – 5000 m
2000 – 3000 m
1000 – 2000 m
500 – 1000 m
200 – 500 m
0 – 200 m
Ice coverage
International boundary
Road
Railway
Airport
Capital city

The launch was a moment of great pride for the Indian government and its 50-year-old space programme, but it also sparked a negative reaction amongst many people at home and around the world due to a feeling that India had got its priorities wrong. Opposition was particularly marked amongst some people in the United Kingdom, which had donated over US$800 million in **aid** contributions to fight poverty in India in 2013.

How could the Indian government possibly justify this annual expense at a time when the World Bank has ranked the standard of living of its people to be a lowly 148th out of 189 (the USA was 13th), when two-thirds of the country's children are officially **malnourished** and half the population lacks modern internal toilets? How can all of this be worth the money?

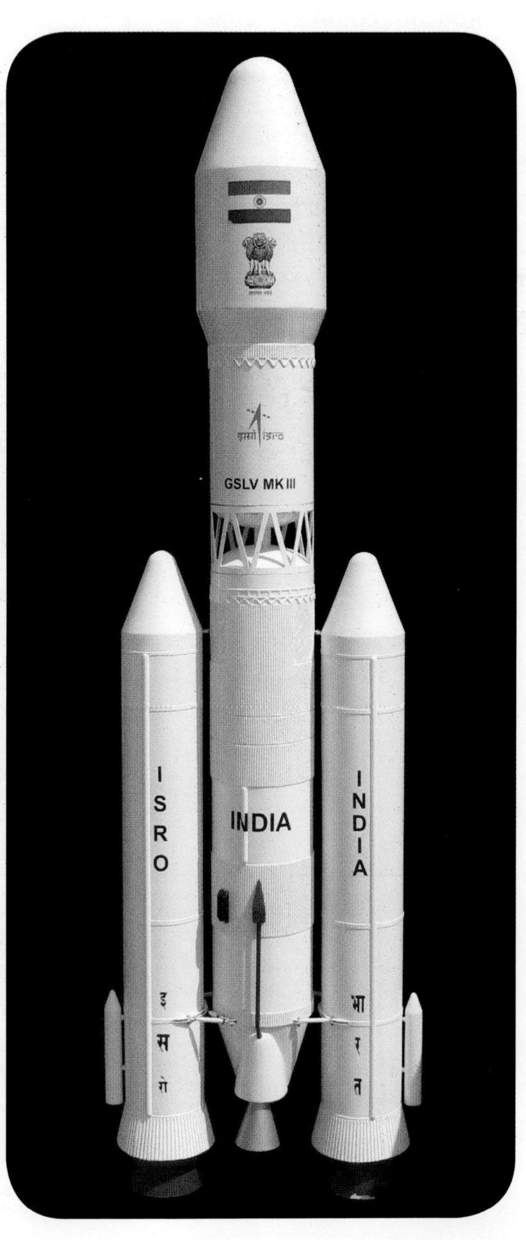

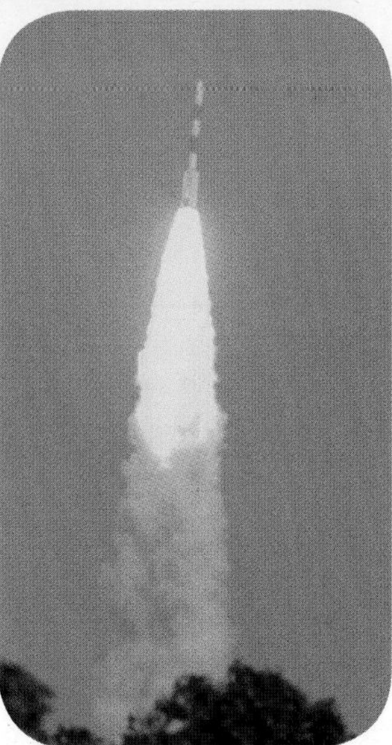

Make a list of as many limitations of GDP per capita as you can discover from the following sources:

http://www.s-cool.co.uk/a-level/geography/world-development/revise-it/measuring-differences-in-development

http://www.tutor2u.net/economics/content/topics/livingstandards/limitations_of_gdp.htm

http://econperspectives.blogspot.co.uk/2008/08/limitations-of-using-gdp-as-measure-of.html

Can you think of a more appropriate indicator which would more accurately reflect differences in the quality of life of people living in different countries around the world?

▶ Consolidating your thinking ◀

The indicator that is regularly used by organisations such as the World Bank to assess the standard of living of a country's population is known as **gross domestic product (GDP)** per capita. This is calculated by adding together the value of all goods and services produced within a nation in a given year, converted to US$, and dividing by the mid-year population of the country in the same year. This results in a figure of US$1499 for India, compared with US$53,143 for the USA and a world average of US$10,472. Most often the figures are presented in a choropleth map, such as the below map on GDP per capita. Although it is a common way of comparing levels of prosperity between one country and another, geographers always warn that this indicator should be used with caution because of its limitations.

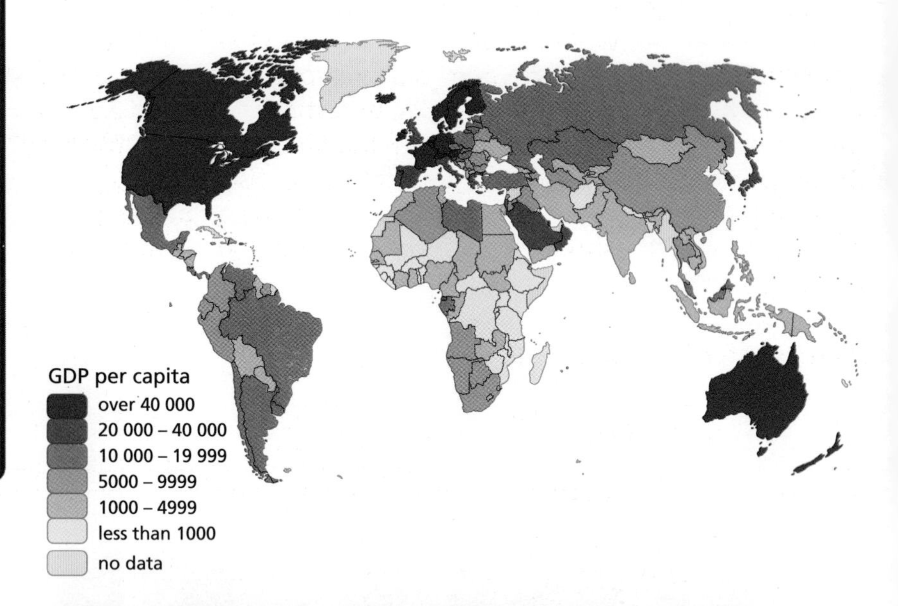

3.1 Why is India's space programme being criticised?

'We can go to Mars but cannot provide clean water for our people on Earth.'

Tavleen Singh,
Indian columnist and political writer
(https://twitter.com/tavleen_singh)

'It's atrocious that taxpayers are still handing money to a country rich enough not just to have its own space programme, but one that is blasting off to Mars. If India can afford this kind of expenditure, then it doesn't need a penny of British taxpayers' money in aid.'

Jonathan Isaby,
Chief Executive of the British TaxPayers' Alliance

Those opposed to continued investment in the Indian space programme and recent projects such as the Mars mission point to the fact that India only has, at best, a medium level of **economic development** and that hundreds of millions of people living there continue to exist in poverty. According to the World Bank's definition of poverty as living on US$1.25 or less a day, there are 270 million people in India coping with poverty. This accounts for 21% of everyone in the world living in poverty based on the same definition.

▶ Consolidating your thinking ◀

Look at the data, images and maps on this page. In terms of quality of life and living standards, how does poverty in India manifest itself? Which of these challenges do you think should be prioritised by the government of India? If solved, which do you feel would have the quickest effect in terms of boosting economic development and prosperity for people in the country? Create a rank order of priorities from most to least urgent and then compare your thinking with a partner. Be prepared to contribute to a whole group discussion. Is there a high degree of agreement amongst the group as to what the priorities should be? If not, why do you think there is disagreement?

Poverty rates in India

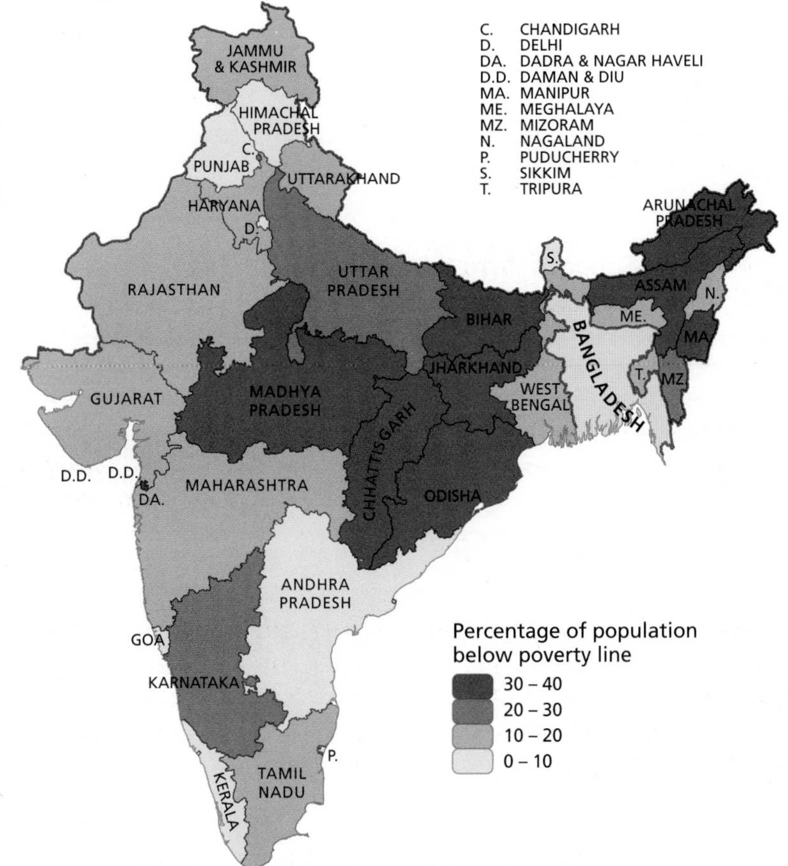

Because of the limitations of GDP per capita as an indicator of development and as a means of comparing the quality of life in one country with another, in recent times geographers have created more relevant and sophisticated measures of standard of living that do not rely solely on economic data. One of these modern indicators is the **Human Development Index (HDI).**

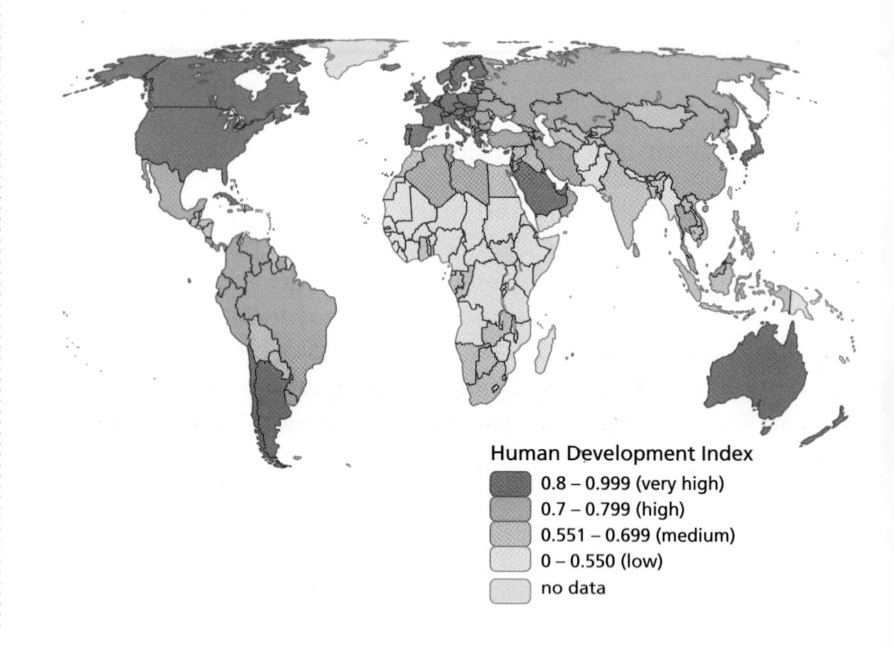

The HDI is considered to be a much more accurate measure of development than GDP per capita because it is a summary measure of *three* key dimensions of human development: a long and healthy life, being knowledgeable and having a decent standard of living. Adopting this measure often means that countries with the same GDP per capita can end up with very different Human Development indices. See page 26 of the Teacher Book for a table showing HDI for India and the USA.

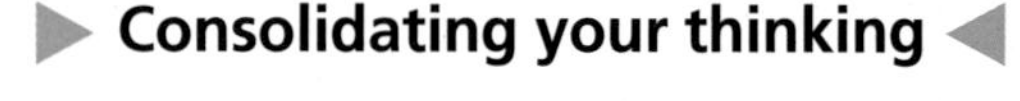

Consolidating your thinking

The region with the lowest average HDI is **Sub-Saharan Africa**, which includes all African countries that are fully or partially located south of the Sahara Desert (excluding Sudan).

Poverty in Sub-Saharan Africa has been rising over time and its countries face a situation worse than most other countries in the world; nearly 50% of the population is classified as poor. Spend some time identifying the reasons which explain why this area is the poorest in the world. Try to categorise the reasons into physical and human causes. Think about whether there are any similarities between the causes of poverty in Sub-Saharan Africa and the causes of poverty amongst many people in India.

Begin your research at:

http://www.eoi.es/blogs/lauraambros/2012/01/17/millenium-development-goals-for-sub-saharan-africa/

http://pages.towson.edu/thompson/courses/Regional/Lectures/SSA/africapoverty.pdf

http://www.worldhunger.org/articles/Learn/africa_hunger_facts.htm

http://en.wikipedia.org/wiki/Poverty_in_Africa

The Teacher Book provides details of how the **Multidimensional Poverty Index (MPI)** is calculated on page 27. Many geographers consider that the MPI provides the most accurate assessment of the standard of living of people in a country.

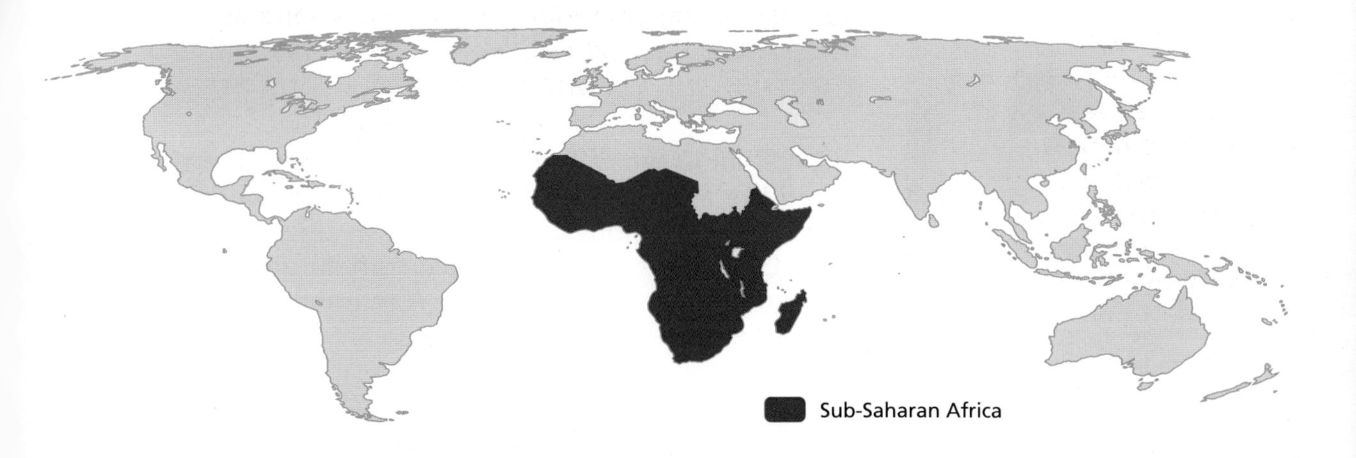

Consolidating your thinking

For this exercise, you will need a copy of both the 'Table of MPI in 107 developing countries' (pp.28–29) and the outline *scatter graph* (p.30) from the Teacher Book. The table shows three variables for each country:

- The percentage of people who are assessed as MPI poor.
- The average intensity of that poverty amongst the people who are MPI poor.
- The total number of people assessed as MPI poor in that country.

Using the outline, construct a scatter graph to prove or disprove the hypothesis that in developing countries there is a positive **correlation** between the percentage of people who are assessed as MPI poor and the intensity of that poverty. In other words, is it true to say that in poor countries, as the percentage of MPI poor people increases, so does the intensity of their poverty?

What you need to do:

- Firstly, plot the 107 points on the scatter graph – one for each country.
- Draw in a line of 'best fit'.
- Decide whether the relationship (or correlation) between the two variables is 'positive', 'negative' or 'no correlation'.
- Describe and try to explain your results – what might be causing the relationship or pattern? Think about this with a partner and then contribute your ideas to a whole group discussion.

3.2 Why do people support India's space programme?

In 2008, India sent a mission to the moon and in the years ahead the Indian Space Research Organisation (ISRO) has even more ambitious plans. As well as the Mars probe in 2013, ISRO plans to send a space craft to Venus in 2015 and then another probe to the sun. A reusable launch vehicle is also being developed.

According to richtop10.com, the top ten most technologically advanced countries in the world in 2014 were:

1. Japan

2. Singapore

3. United States of America

4. United Kingdom

5. Canada

6. France

7. China

8. Sweden

9. Australia

10. Finland

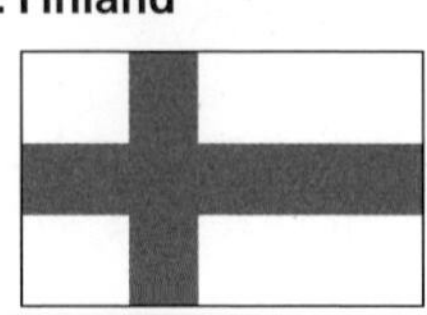

'These achievements, however, haven't stopped detractors from asking why India is doing this when a third of its people live below the international poverty line. The simple answer is because it makes economic sense, as technological and social development goes hand in hand. This reasoning has been embraced throughout the countries of the developing world.'

Akshat Rathi, journalist, 'Poor countries want space programmes more than rich ones do', 2011

'Capturing and igniting the young minds of India and across the globe will be a major return for this mission.'

P. Kunhikrishnan, Chairman of ISRO and 2013 Mars Mission Director

Consolidating your thinking

Using a copy of the table in the Teacher Book titled 'Standard of living in the top ten most technologically advanced countries in the world' (p.31), fill in the relevant ranking of each country for:

- GDP per head ranking according to the International Monetary Fund, World Bank, CIA World Factbook and the United Nations at: http://en.wikipedia.org/wiki/List_of_countries_by_GDP_(nominal)_per_capita
- Human Development Index ranking at: http://en.wikipedia.org/wiki/List_of_countries_by_Human_Development_Index
- Organisation for Economic Co-operation and Development (OECD) and International Labour Organisation average wage (adjusted to reflect relative cost of living) ranking from: http://en.wikipedia.org/wiki/List_of_countries_by_average_wage
- Happiness Index ranking at: http://en.wikipedia.org/wiki/Happy_Planet_Index

What does your completed spreadsheet tell you? Is it generally true that the most technologically advanced countries in the world are also the countries in which people enjoy the highest standards of living? Which of these ten countries has the highest rank for the first seven economic indicators of development and which the lowest? Which country in the top ten is the *exception* to the general rule that technological development brings high standards of living? Why do you think this is?

Read the articles at:

http://www.forbes.com/sites/timworstall/2013/09/05/apple-gets-it-in-the-neck-over-chinese-labour-standards-again/

http://www.facing-finance.org/en/database/cases/working-conditions-in-foxconn-factories-in-china/

and watch the news reports at:

http://www.youtube.com/watch?v=yQPrbwWWUD4

http://www.youtube.com/watch?v=h1_XAuJ8qCc

What allegations do these reports and films make about the conditions of many factory workers in China employed by companies based in America and Europe?

World happiness

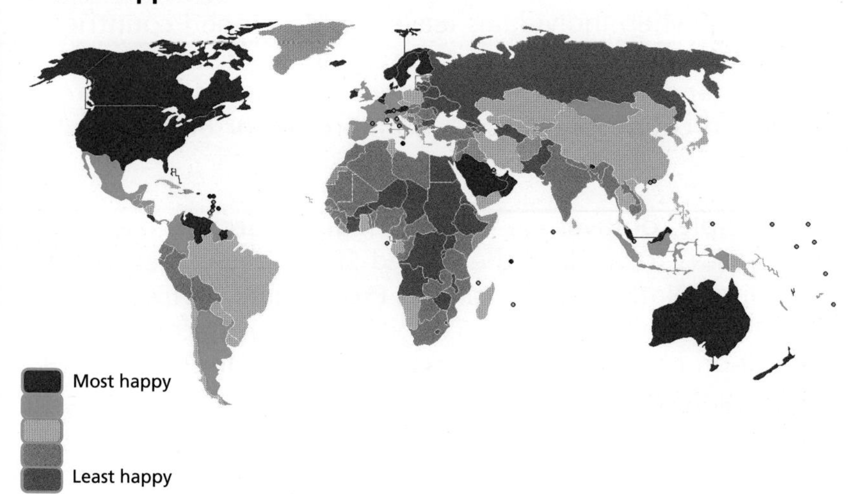

To the left is the first map ever compiled of world happiness. It was created by Adrian White at the University of Leicester and is the most radical alternative to using GDP per capita as an indicator of people's quality of life.

Supporters of India's space programme use the **multiplier effect** as a strong argument for continuing to invest money in space exploration. They point out that the 16,000 jobs that exist in the ISRO (mainly engineers and scientists) have led to at least 100,000 new jobs being created already.

These new positions are in businesses and services around India which either manufacture goods that the programme needs, such as circuit boards, or provide services like security and catering. In this case the multiplier effect is 1:5, meaning every new job in the space programme leads to five others being created.

Consolidating your thinking

Using a copy of the multiplier effect diagram in the Teacher Book (p.32), write a paragraph to explain why investment in India's space programme headquarters can become more and more popular as a location for new business over time. Write a paragraph to explain what would happen to the area if one major business, such as the space programme, either declined or closed down altogether.

Many people in India argue that investing in a space programme and employing the best young scientists coming out if its schools and universities is essential to slow down the country's **brain drain**.

The term *brain drain* refers to the **emigration** (out-migration) of knowledgeable, well-educated and skilled professionals from their home country to another country. The main reason for this is the availability of better job opportunities in the new country. Brain drain occurs most commonly when individuals leave less developed countries (LDCs) with fewer opportunities for career advancement, research, and academic employment and migrate to more developed countries (MDCs) with more opportunities.

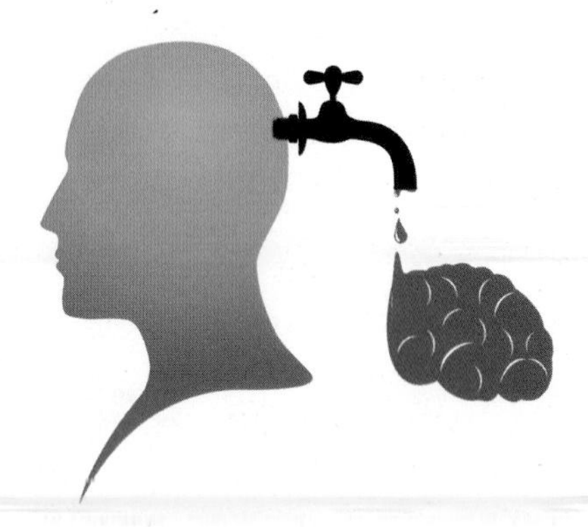

For example, on average 120,762 migrants from India receive visas to work in the United States every year and the United Nations Development Programme (UNDP) estimates that India loses US$2 billion a year because of the emigration of computer experts to the USA. In 2012, over 3000 Indian doctors went abroad for studies and did not return.

The brain drain of skilled men and women is most marked in the countries of Sub-Saharan Africa which, as you know, has the lowest HDI of any region – in other words it's the poorest part of the world. From the map of the region on page 43, select any one of the countries shown and find out where most university-educated people who leave go and what they do when they get there.

There are many things governments can do to combat brain drain. According to the *OECD Observer*, *'Science and technology policies are key in this regard. The most beneficial tactic would be to increase job advancement opportunities and research opportunities in order to reduce the initial loss of brain drain as well as encourage highly-skilled workers both inside and outside the country.'*

Make a list of the losses or costs of brain drain to countries like India. Which of these do you think is likely to have the greatest impact on the 'donor' country? Similarly, how do you feel a 'recipient' country of brain drain movement of people will benefit from receiving them?

Begin your research at:

http://geography.about.com/od/urbaneconomicgeography/a/braindrain.htm

http://www.migrationpolicy.org/ article/reassessing-impacts-brain-drain-developing-countries

http://www.youtube.com/watch?v=3AnX6rZv86o

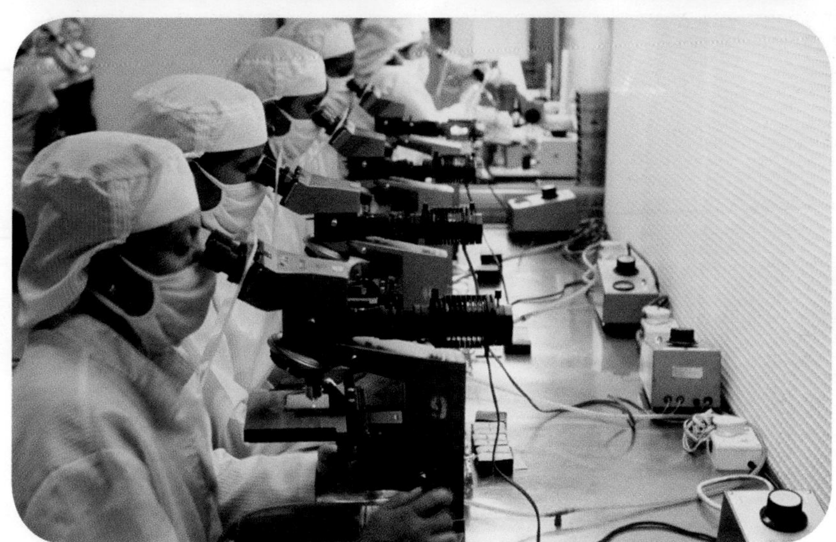

In October 1999, a powerful **cyclone** formed over the Malay Peninsula, 1600 km to the east of India. During the following four days, Cyclone 05B moved west across the north Indian Ocean to become the worst storm in India for thirty years. It was tracked by satellite using the Saffir-Simpson Hurricane Scale. The Orissa region on India's east coast took the brunt of the cyclone as it made landfall, killing more than 10,000 people and causing widespread destruction.

In 2013, the even more powerful Cyclone Phailin hit the same region of India but caused only fifty-one deaths amongst the population. One reason for this was that in the intervening fourteen years the Odisha State Disaster Management Authority had built 200 new cyclone shelters.

However, the main reason for such a small loss of life was that India's space programme had been able to develop and launch a number of advanced weather satellites to closely

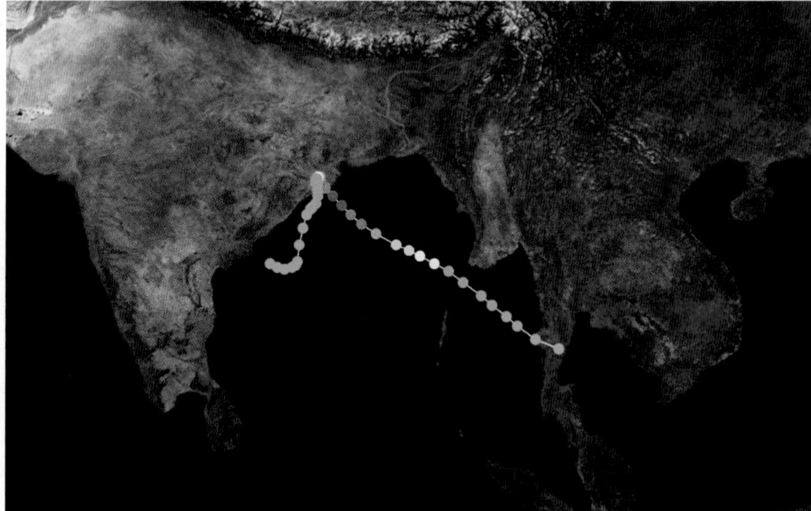

Saffir-Simpson Scale

CATEGORY 1	CATEGORY 2	CATEGORY 3	CATEGORY 4	CATEGORY 5
Minimal damage Wind (kph) 119-153 Surge (m) 1.2-1.5	Moderate damage Wind (kph) 154-177 Surge (m) 1.8-2.4	Extensive damage Wind (kph) 182-209 Surge (m) 2.7-3.7	Extreme damage Wind (kph) 211-249 Surge (m) 4.0-5.5	Catastrophic damage Wind (kph) >249 Surge (m) >5.5

monitor the movements of cyclones in the Indian Ocean. This is another benefit of space technology and is often highlighted by the Indian government and supporters of continued investment in the programme.

The government of India believes that anyone who doubts the benefits that a space programme can bring to their country should look no further than the city of Bangalore, the headquarters of the ISRO since 1969, now referred to as the '**Silicon Valley** of India'. Since then, the city has become the centre of India's defence industry, a huge technology park has been developed and a massive IT services industry has built up around the space programme. Today, both home grown technology giants such as Infosys and global multinational business software firms including Microsoft, IBM and SAP make up the 500 IT companies that have centres in the city. Over 40% of India's IT industry is to be found in Bangalore and together these firms bring in US$17 billion a year for the country.

According to the Saffir-Simpson Hurricane Scale, how strong were the wind speeds when Cyclone 05B made landfall in 1999?

Read the United Nations Development Programme (UNDP) report at:

http://www.unep.org/pdf/UNEP_GEAS_NOV_2013.pdf

about how satellite early warning systems have reduced India's vulnerability to potential **natural disasters**. What did the early warnings allow people living in coastal districts to do before Cyclone Phailin hit land?

▶ Consolidating your thinking ◀

Watch the film and read the BBC News report, 'Bangalore: India's IT hub readies for the digital age' at http://www.bbc.co.uk/news/technology-23931499. What is the main advantage of the IT industry in Bangalore compared with locations in the United States, such as Silicon Valley in California? Do you think that India will always have this advantage as it develops economically and the standard of living of people rises in the future? What might happen then?

Those in India who want to see the country's space programme expand in the future also highlight that at a cost of US$1 billion a year, it is incredibly cheap compared with those of other countries, such as the United States, Russia and China, and worthwhile in terms of what the country gets back in return from a relatively tiny investment. They say the US$1 billion spent on developing the space industry is only a pittance compared with:

- US$20 billion spent each year providing cheap food for about 800 million Indians including a school feeding programme providing a meal for all children every day.
- US$5.3 billion spent each year on a rural employment plan.
- US$3.5 billion which India gives in aid contributions to other countries.
- US$24 billion which India contributed to the International Monetary Fund to help the countries of Europe cope with the **'Eurozone crisis'**, 2010–2013.

'India is quite capable of sending a rocket to Mars and fighting poverty at the same time. In fact the space programme has so far been a hero of India's pursuit of development. Its annual budget is just 0.038% of India's GDP. The direct and indirect benefits of this minuscule cost to the country are huge and are contributing to lifting people out of poverty by generating jobs and paying decent wages.

'Our space rockets have already launched 40 satellites for foreign countries that pay us millions of dollars because our programme is efficient and reliable. If people in this country are to improve their lives then a successful and growing space programme that brings in money and creates jobs is not a luxury but essential. The benefits are many, but most Indians probably do not realise how their space programme touches them. All of this for a tiny fraction of government spending – seems like a no-brainer to any policy maker.'

Indian online blogger, 2014

Watch the US Public Broadcast Service interview with Dr Radhakrishnan at:

http://www.pbs.org/newshour/bb/world-jan-june14-indiaspace-01-18/

and read the transcript. Thinking back over your investigation so far, are you inclined or disinclined to agree with Dr Radhakrishnan when he says that 'the space industry is touching the lives of every man and woman in this country'?

Today, India has one of the fastest growing economies in the world. In 2007, only 250 million people were in the more wealthy 'middle class'. This is set to increase to 600 million (almost one in three people) by 2030. India's space programme is one of the reasons for this. Go to Gapminder at www.gapminder.org and play the chart for India. In terms of income and life expectancy, in which year would you say India began to increase rapidly? How does this date compare with the development of the country's space programme?

3.3 What can India learn from the experience of the USA's space programme?

The **National Aeronautics and Space Administration (NASA)** is the government agency in the USA responsible for the civilian space programme as well as for aeronautics and aerospace research. The agency became operational in 1958 and since that time, most of the USA's space exploration efforts have been led by NASA, including the Apollo moon-landing missions, the Skylab space station, and later the Space Shuttle and International Space Station. NASA's budget has generally been approximately 1% of the federal budget from the early 1970s onwards, but briefly peaked to around 4.41% in 1966 during the Apollo programme. Currently, it is just below 0.5% of the US$3.77 trillion government budget expenditure, amounting to US$17.8 billion a year.

On pages 52 to 53 are a range of NASA 'spin offs' – things that were originally developed as part of the USA's space programme but now have applications which benefit ordinary people in the world and the environment on which they depend.

NASA SPIN-OFFS

Working with a partner, discuss why you think the USA's space programme has needed to develop each of these technologies and how their application in ordinary life today is a benefit to people and/or the environment and America of course, which sells this technology to companies all around the world!

Cordless tools and appliances

Solar energy

The development of personal computers

Managing the spread of cities

Wildlife conservation

Polymers advanced heat management materials for vehicles

River flood control

Using ions to sterilise drinking water

Managing ocean pollution

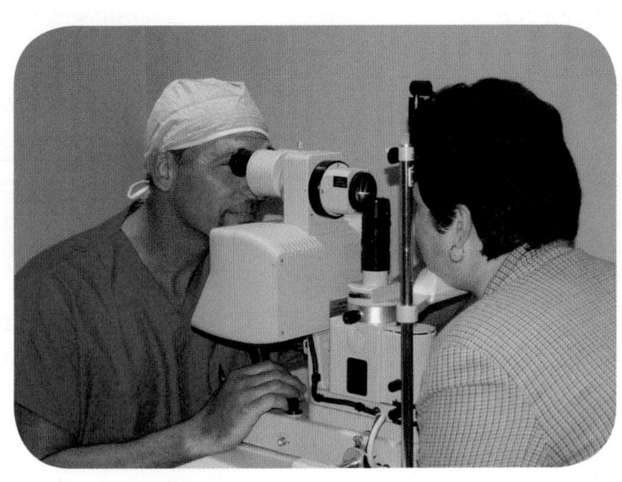

Laser eye surgery

Applying your skills

In 2013, *The Economist* news magazine reported that:

'India is not the only emerging economy with space ambitions. Nigeria already has a handful of satellites floating around the Earth (though these were launched by others). Depending how you define a space programme, even minnows like Sri Lanka, Bolivia and Belarus have plans of some sort to get space activity under way. By one count, including co-operative efforts between countries but not fully private ones, there are currently over 70 space programmes, though only a dozen of these have any sort of launch capability.'

See more at http://www.economist.com/blogs/economist-explains/2013/11/economist-explains-0#sthash.is28XPQq.dpuf.

Working with a partner, choose one of the emerging space nations on the map on page 33 of the Teacher Book. The government of this country is preparing to spend over US$1 billion dollars to grow its space programme so that it has its own launch spacecraft capability by 2020.

They are anxious that people support them. In order to convince everyone of the value of developing their space programme, they are looking to hire an advertising agency to produce a two-minute television advertisement to be shown at peak viewing times every day for a week.

Your task is to plan, create and produce this advertisement. Globally, over US$50 billion a year is spent on television advertising and the government of the country you are now representing is aware of how much influence it can have on people's views. One single company, Proctor and Gamble, spends US$5 billion on advertising each year.

Filmmakers, animators and web developers all use storyboarding to sketch out and organise the running order of scenes and narrative before filming begins and to ensure that the story with the key messages flows and makes the very best use of time. Filmmaking is very expensive! There is an example in the Teacher Book (p.34) which you can use as a model to help you lay out your storyboard.

Using copies of the storyboard template in the Teacher Book (p.34), plan and create the advertisement. Within your design, use what you have learned about the benefits of the space programmes in India to convey key messages. What are the things you are going to emphasise in your advertisement and what images will you choose to go with the narration? How are you going to address the concerns that some people will inevitably have about spending this amount of money?

By 2014, at least fifty-two nations around the world were already operating one or more satellites that have been carried into space on rockets belonging to other nations such as India. Collectively these countries are known as *emerging space nations* and a selection of twenty-seven are located on the world map on page 33 of the Teacher Book.

An advertising agency will do a lot of research about both the product they are promoting (you have done that already) and the *place* in which and *the people to whom* the advertisement is going to be shown. Therefore, you will need to spend enough time researching in depth the country of the government that is commissioning you.

For example, you will certainly want to know the country's current level of economic development and the quality of life of the people that live there in order to pitch the advertisement correctly. You will also want to know what potential the country might have to take advantage of a space programme. For example, does it have natural resources that are currently not being exploited because it lacks the technology to develop them?

In addition, make sure that you are aware of any challenges currently facing the country, such as insufficient food growth to feed its population or environmental problems like **desertification** or the effects of **climate change**. If you have a good understanding of these then you can address them in the advertisement.

Finally there is the 'pitch'. The pitch states the objectives for the advertisement and describes how it will deliver its intended result – to persuade people to support the government's proposal to develop an expensive space programme.

You will need to run through your storyboards and describe how the advertisement will unfold and the key messages you will make on their behalf. The rest of your group will play the role of the government officials that you need to convince.

Good luck!

Geographical Enquiry

4 From Shangombo to South Tavy Head

Why are Thandi and Moses working in Soweto market?

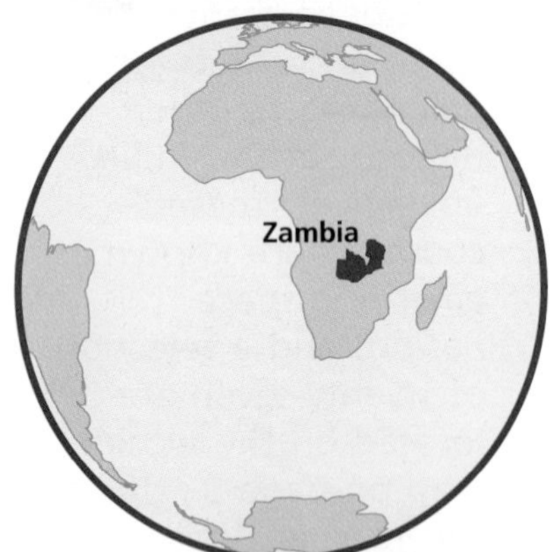

Fifteen year old Thandi and her younger brother Moses work seven days a week in Soweto market in Lusaka, the capital and largest city in Zambia in southern Africa. They survive by buying produce such as bananas and eggs from local farmers and selling them on for a few kwacha of profit at the market. They left their home, mother and four brothers and sisters in the Shangombo District of Western Zambia a year ago and now live with their aunt. They have not been home since. Like 43% of primary school aged children (7–13 years) and 78% of secondary school aged young people (14–17 years) in Zambia, they don't attend school.

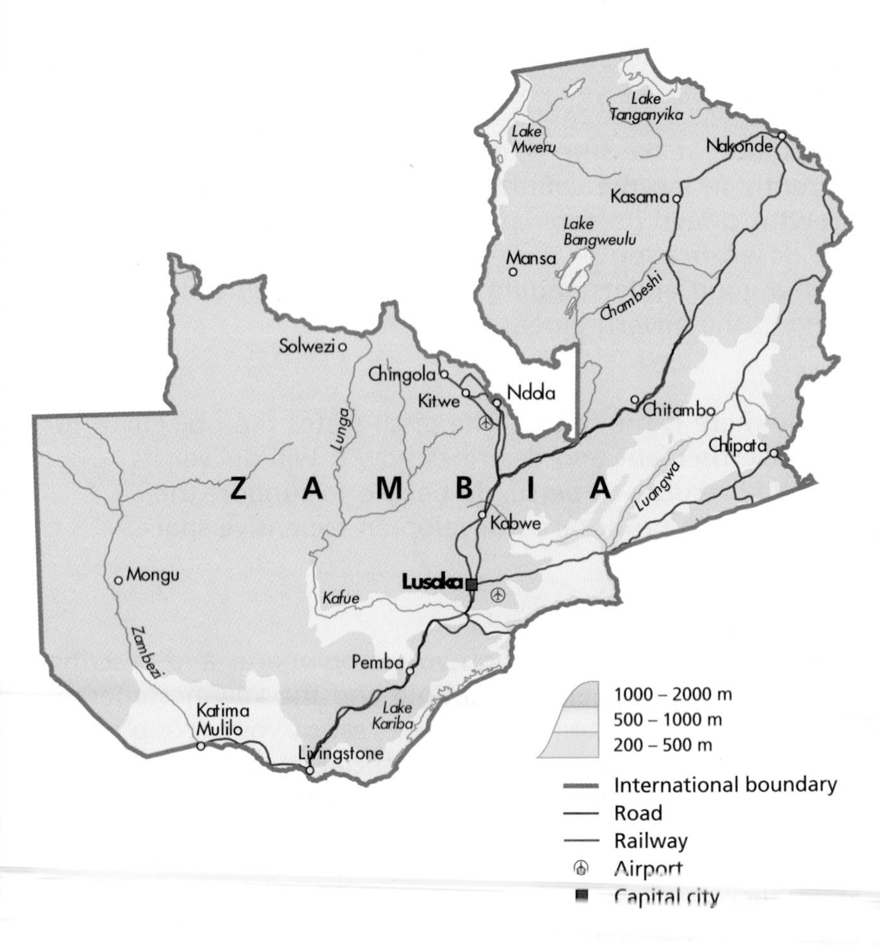

4.1 What is the geography of Shangombo like?

Shangombo is one of the most geographically remote districts of Zambia. It lies close to the border of Angola and is a fourteen-hour journey from Lusaka, along a largely untarmacked road.

Consolidating your thinking

Look at the images and information on these pages and at the table of selected poverty related indicators in the Teacher Book (p.38). What do they suggest to you about the physical, social and economic geography of Shangombo? Use a copy of the template in the Teacher Book (p.39) to record your ideas.

The normal climate pattern of the area presents an annual challenge to farmers. From the data in the table below, construct a climate graph for Shangombo.

From a farmer's perspective, what two challenges does the rainfall pattern present in particular? Read the three quotations about environmental problems in Shangombo in the Teacher Book (p.40) and study the images to clarify your thoughts.

4.2 How is the environment a constraint to living in Shangombo?

Shangombo is one of the hardest places in Zambia in which to live. One of the reasons for this is its geographical remoteness. For example, it is not yet connected to the national electricity power grid of Zambia. The majority of the population depend on farming – either rearing cattle to sell at markets further east or as **subsistence farmers** of maize, millet and vegetables.

Climate data for Shangombo

Shangombo	Jan	Feb	Mar	Apr	May	Jun	Jul	Aug	Sep	Oct	Nov	Dec
Temperature – max. (°C)	29	29	29	30	28	27	27	30	33	34	31	29
Temperature – min. (°C)	19	19	18	17	13	10	10	12	16	18	18	19
Rainfall (mm)	209	185	140	43	5	1	0	2	2	33	106	193

Over the past twenty years, weather patterns have changed dramatically throughout the countries of southern Africa. Farmers who have known for centuries when to expect the rainy season are now finding planning difficult. It is becoming impossible to predict year on year what the pattern of rainfall will be. In 2013, UNICEF, the United Nations' children's agency, declared that the western districts of Zambia were experiencing their worst **drought** in thirty years, leading to the widespread death of livestock and the failure of crops. Farmers in Shangombo now have to rely on aid shipments of maize seed from their own government and from overseas.

Surface air temperature change (°F)
(2050s average minus 1971-2000 average)

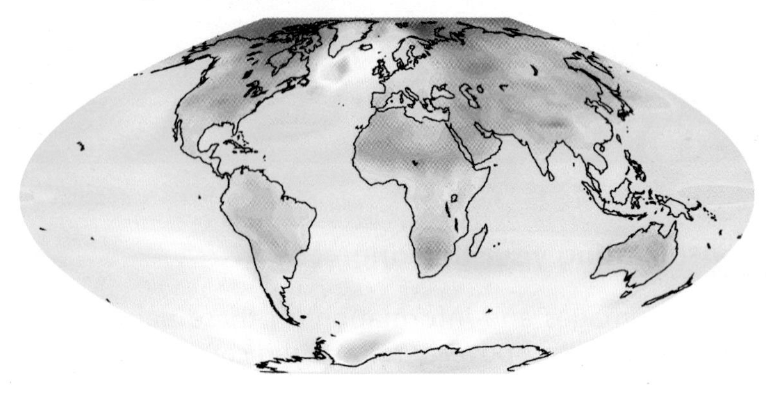

Consolidating your thinking

Once the annual rains fail or are less than normal, a 'vicious circle' of soil and land degradation begins. Rearrange the stages in land degradation below into the correct order and on a large piece of paper create a circular flow diagram with annotated sketches or photographs to illustrate each stage. Remember that this is a circular system, which means that an arrow should link the last stage back to the first. Use the example diagram on page 41 of the Teacher Book to help you.

- Wind **erosion** removes exposed soil.
- Cattle die and crops fail.
- **Pluvial flooding** removes soil.
- Rains fail.
- Vegetation dies and ceases to anchor the soil in place.
- Local people depend on aid shipments to survive.
- Occasional torrential downpours cause **flash floods**.
- Food shortages.
- Soil becomes increasingly exposed to the elements.
- Vegetation cover begins to die back.

As well as drought, other human factors often make the effects of land degradation worse. Discuss the images on these pages and decide how they contribute to land degradation and add them to your diagram.

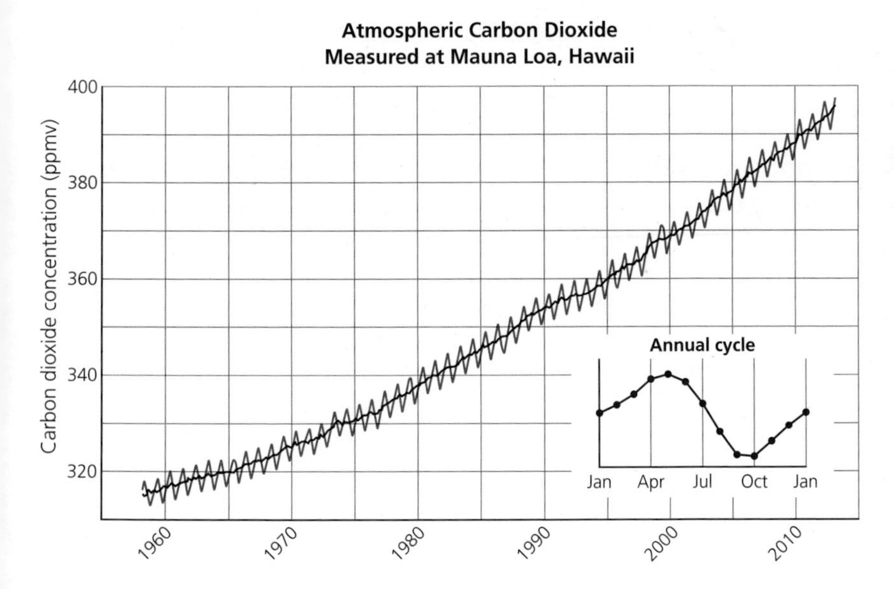

4.3 Why are the world's drylands so important?

The Shangombo District of Western Zambia is part of the world's drylands, which are shown on the map below. As their name suggests, drylands are precisely that – dry. They all cover places like Shangombo that receive low annual amounts of **precipitation** in the form of rainfall or snow. More precisely, drylands are defined by the United Nations (UN) as having an **aridity index** of 0.65 or less.

World aridity

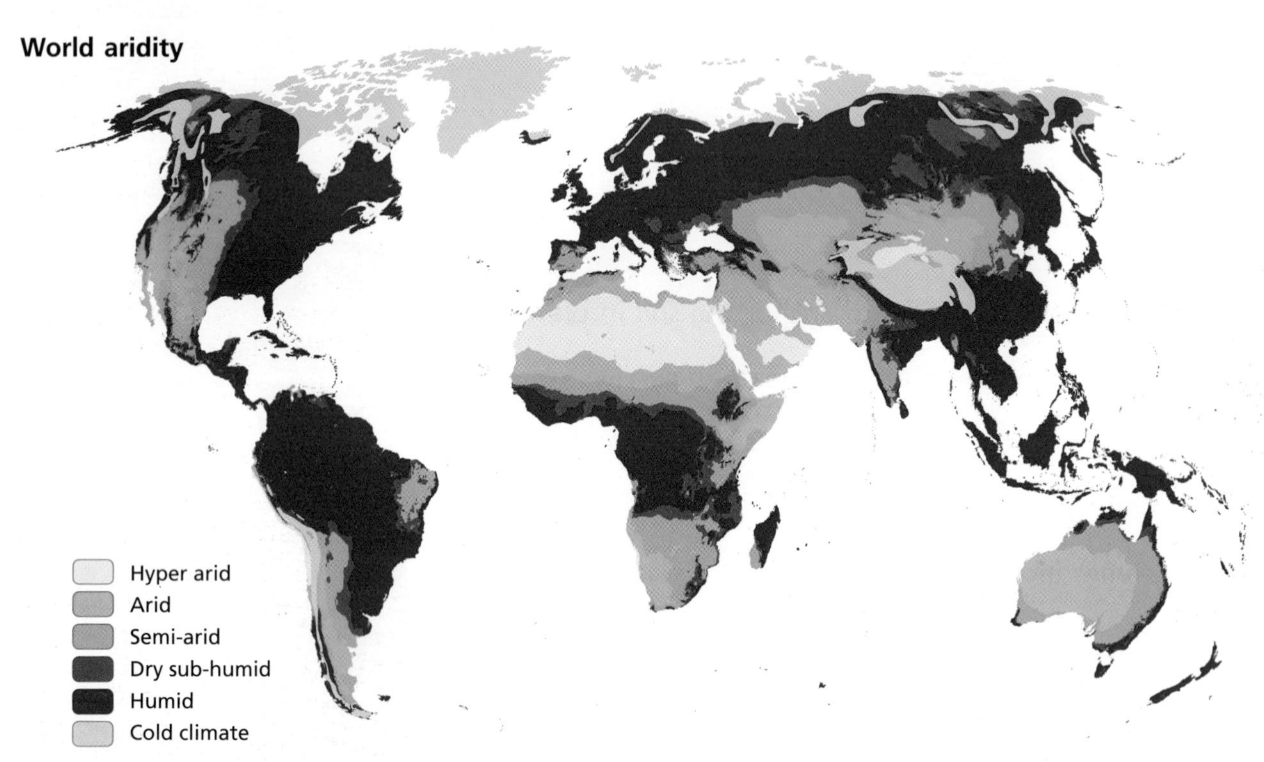

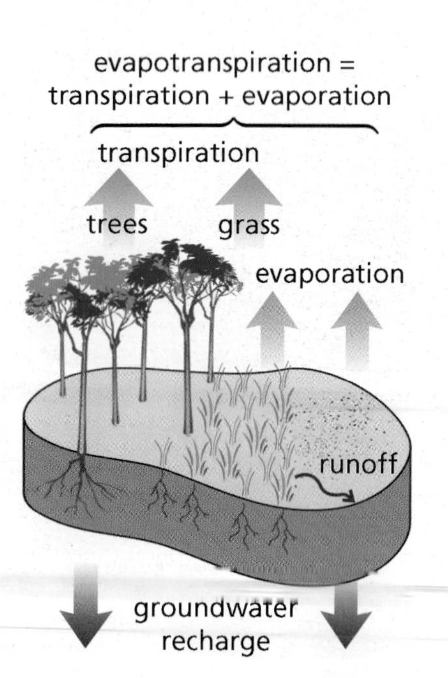

The aridity index is a measure of the ratio between average annual precipitation and total annual potential **evapotranspiration**. It is calculated by dividing the average annual precipitation (mm) of a place by the potential evapotranspiration of that place (also in mm). Evapotranspiration is a measure of the amount of water which returns to the **atmosphere** as a result of evaporation from the land surface (including soil) and from vegetation, such as trees.

In almost every place in the world this calculation will generate a number between 0 and 3. The lower the number, the more **arid** or dry the place is. There are four categories of dryland in the world:

Category	Index	% share of world's land area	% share of world's population	% of area used for pastoralism	% of area used for cultivation	Additional notes
Hyper-arid	0–0.05	6.6	1.7	97	0.6	Rainfall is less than 100 mm a year. Droughts of longer than a year are common. Only nomadic pastoralism is possible. Almost all subsistence.
Arid	0.05–0.20	10.6	4.1	87	7	Generally rainfall does not exceed 200 mm a year. Some irrigated crops are grown but mostly extensive livestock farming. Largely subsistence.
Semi-arid	0.20–0.50	15.2	14.4	54	35	Rainfall does not exceed 800 mm a year. Mostly mixed cattle rearing with staple arable crops.
Dry sub-humid	0.50–0.65	8.7	15.3	34	47	Sufficient rain for extensive commercial crop growing but rainfall is highly seasonal and long dry periods can occur.
Total		41.3	35.5	65	25	

Table showing the four dry categories of the aridity index

Consolidating your thinking

One in three people live in dryland areas, which cover 41% of the world's land surface. Furthermore, a much larger proportion of drylands are in developing countries (72%) and this proportion increases with the level of aridity. All of the hyper-arid drylands are in the the poorest regions of the world. Although mostly uninhabited, there are small communities who live year-round in the world's driest places.

Over one billion people rely directly on the dryland ecosystem for their daily survival, mainly through rain-fed or **irrigated** farming, or through extensive **pastoralism**. Amongst these peoples are small **indigenous** groups, whose way of life and relationship with the environment that supports them has remained largely unchanged for thousands of years.

Choose one of the following groups and research all of the ways in which they are **adapted** to living in such a harsh environment: *Bedouin*; *Australian Aborigines*; *Tuareg*; *San Bushmen*; *Kyrgyz.* In what ways could their way of life still be described as **sustainable**?

Some of the world's biggest cities such as Cairo and New Delhi are found in the world's drylands. One of these cities is Las Vegas in Clark County, Nevada in the USA. Established in 1905 in the Mohave Desert, today the city has a population of 600,000, with two million people living in the wider Las Vegas metropolitan area. Its population increased by 66% during the decade 1990 to 2000 alone, which made it America's fastest expanding **urban** area. The residents, together with the forty million tourists who fill the city's hotels each year, are almost entirely dependent on the Colorado River basin for their water. Only 15% of the water the city requires is extracted from underground **aquifers**. Just outside the city, the Hoover Dam was constructed between 1931 and 1936 to create Lake Mead in order to supply Las Vegas' water needs.

Today, the city of Las Vegas faces a major water supply challenge which could result in it becoming just another abandoned desert 'ghost town' in the future. Read the following extract from a report on CBS News, broadcast in January 2014. What is the problem and what is causing it?

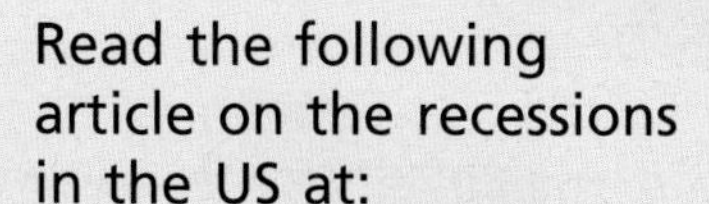

Read the following article on the recessions in the US at:

http://www.bloomberg.com/apps/news?pid=newsarchive&sid=aNivTjr852TI

'When you head out on Nevada's Lake Mead, the first thing you notice is a white line. That's where the water used to be.

'What did this look like a decade ago?

'"This was all underwater," said Pat Mulroy, the general manager of the Southern Nevada Water Authority. "I mean boats were everywhere. There was a whole marina here."

'Mulroy said that the drought began 14 years ago. Satellite photos show the Colorado River, which feeds Lake Mead, is drying up – so the lake is rapidly shrinking. Islands are growing, and boats are floating far from where they once were.

'"It's a pretty critical point," Mulroy said. "The rate at which our weather patterns are changing is so dramatic that our ability to adapt to it is really crippled."

'The bathtub ring is going to get bigger. Lake Mead is expected to drop at least another 20 feet this year. If it does that could trigger automatic cuts to the water supply for Nevada and Arizona.'

CBS News, 30 January 2014

Consolidating your thinking

In response to this challenge, the Las Vegas legislature has adopted a sustainability plan to reduce its ecological footprint by 30% by 2023. From the sources listed to the right identify what Las Vegas is doing to make the city more sustainable in the areas of air quality, transportation, water stewardship, waste management and recycling and energy efficiency. Do you feel that Las Vegas could be a sustainable community in the future?

Web links to help with your research:

http://www.theguardian.com/sustainable-business/las-vegas-sin-city-sustainable

http://www.lasvegasnevada.gov/sustaininglasvegas/

http://www.cnbc.com/id/38149988#

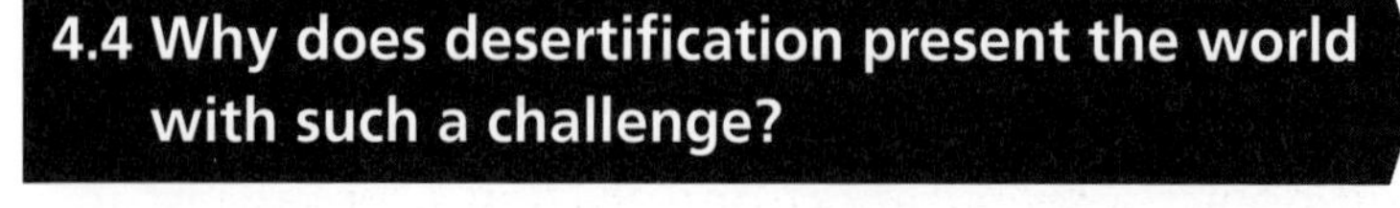

4.4 Why does desertification present the world with such a challenge?

Geographers recognise the process of **desertification** as one of the greatest environmental and development challenges confronting the world in the twenty-first century. The UN estimates that as a result of desertification, twenty-four billion tonnes of fertile top soil disappears from around the world every year – soil which could have supported the growing of twenty million tonnes of grain.

The UN also projects that increased desertification over the next twenty-five years may reduce global food production by up to 12% whilst the population of the world will increase by at least 1.5 billion in the same time interval. Desertification is a root cause of poverty amongst people in many developing countries such as Zambia; because the land in places like Shangombo has undergone desertification, it can no longer support as many people as it previously did. As a result, the out-**migration** of young people to scrape a living in nearby towns and cities is common.

There are several causes of desertification but the most common and widespread is illustrated in the photographs on this page and the next. Look carefully at the images. How and why is the land converting to desert? The photographs show two ways in which the decisions people have made as to how the land is managed has sped up the process of desertification. Can you identify them?

In total, 110 countries across six continents are vulnerable to desertification and its effects. Using the map on page 66 and a political map of the world in an atlas, identify one country in each continent with a very high level of desertification vulnerability – Spain in Europe, for example. For each country you have selected, attempt to discover what the main causes of desertification are and what, if anything, the government of that country is trying to do in order to slow down or halt the land degradation process.

*'Desertification occurs when the tree and plant cover that binds the soil is removed. It occurs when trees and bushes are stripped away for fuelwood and timber, or to clear land for **cultivation**. It occurs when animals eat away grasses and erode topsoil with their hooves. It occurs when intensive farming depletes the nutrients in the soil. Wind and water erosion aggravate the damage, carrying away topsoil and leaving behind a highly infertile mix of dust and sand. It is the combination of these factors that transforms degraded land into desert.'*

International Fund for Agricultural Development (UN agency), 2010

For Spain, your research would take you to several websites such as:

www.climateadaptation.eu/spain/desertification/

www.unesco.org/mab/doc/ekocd/spain.html

news.bbc.co.uk/2/hi/europe/3621228.stm

Overgrazing is the main cause of desertification worldwide. For thousands of years people living in the dryland areas of the world moved with their small herds of domestic cattle, sheep and goats in search of better grazing and water availability. This **nomadic** lifestyle stopped the land at any one location from being put under too much pressure. In modern times the use of fences to control much larger herds has prevented animals moving in search of food and water. This has resulted in overgrazing and desertification. The situation has been made worse through the construction of **boreholes** and windmills at specific locations to bring water from underground aquifers to the surface to create **waterholes**. This allows animals to stay all year in places that were only previously grazed for just a few months during the rainy season. Consequently, they gather around the borehole and overgraze the surrounding area.

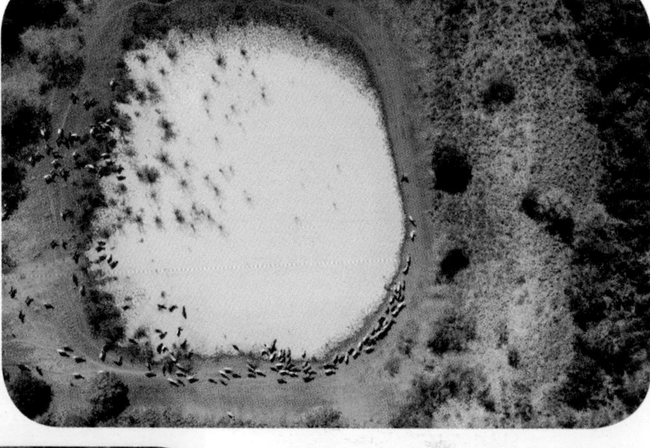

Consolidating your thinking

The average length of a television news report is less than three minutes. Working in a group of four you are going to write the script and then present a three-minute 'news package', as it is often referred to in the media, about Thandi and Moses. The report needs to be filmed on location from the Soweto market in Lusaka, Zambia where Thandi and Moses work. From here, the reporter will talk to camera – this is called a 'standup'. On page 42 of the Teacher Book you will find details of what should appear in your report together with an example page of a completed script and a script template (pp.43–44).

The organisation of a news report follows a clear conventional structure which is explained at:

http://howikis.com/Write_a_TV_News_Script

http://news.bbc.co.uk/1/hi/school_report/6180944.stm

In addition, there are many examples of news reports produced by schools for the 2013 BBC School Report Competition on YouTube to look at for ideas:

http://www.youtube.com/watch?v=l2slmKWpldM

U.S. Department of Agriculture
Natural Resources Conservation Service
Soil Survey Division
World Soil Resources

Desertification vulnerability

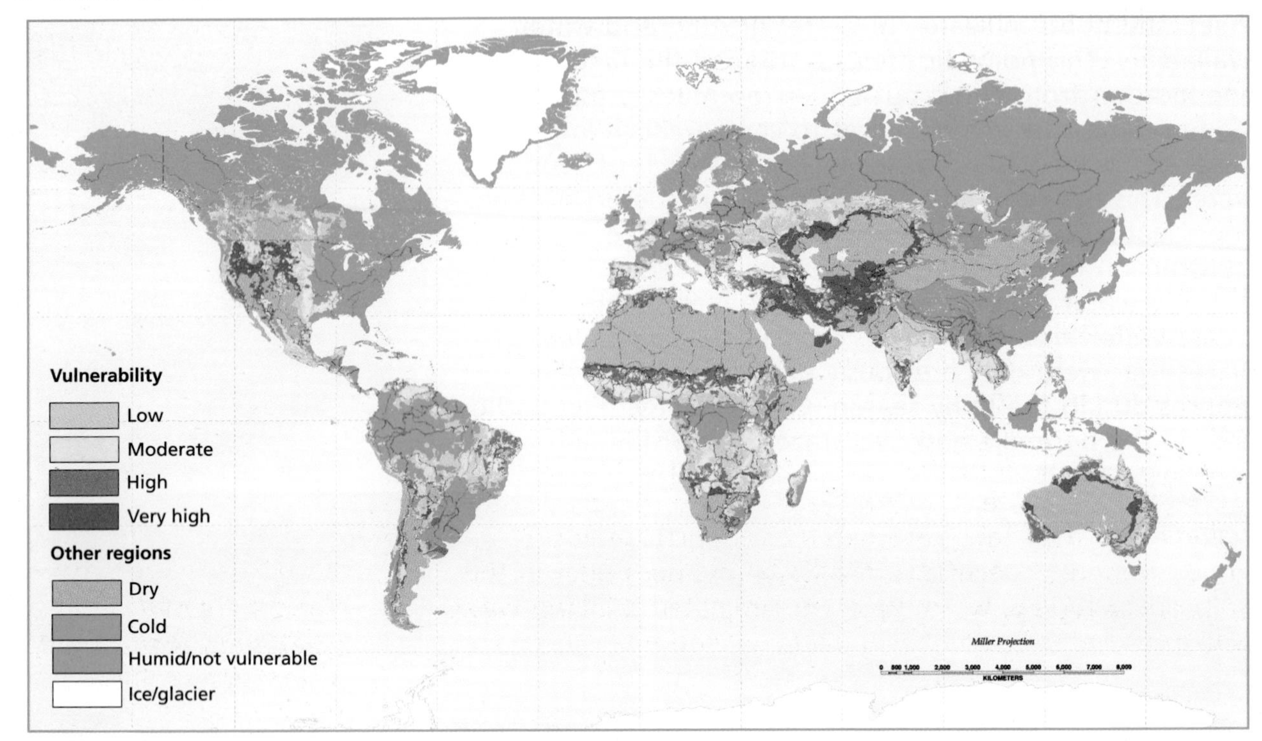

4.5 How are countries around the world restoring degraded land and combatting desertification?

In 1994, many governments around the world signed up to the United Nations Convention to Combat Desertification (UNCCD) and committed themselves to taking action at a local level to restore degraded land and prevent desertification. Rather than encouraging **high-tech** solutions to the problem of desertification, the UNCCD urged countries to use the knowledge and experience of land management possessed by **indigenous** peoples.

▶ Consolidating your thinking ◀

Look at the photographs to the left of farmers in Mozambique and South Africa. Working with a partner, discuss what the images show the farmers doing. One photograph shows *mulching* and the other method is known as *trench and coke bottle farming.* Draw a diagram for each method to show how you think it can **conserve** water and help to restore degraded land. Discuss your reasoning with a partner and then with the group as a whole.

Individual governments with degraded lands all around the world are now taking action on desertification. For example, the governments of Uzbekistan and Kazakhstan are working together to try to restore the devastation caused to the Aral Sea during the 1960s and 1970s when the area was part of the Soviet Union. The drying up of the Aral Sea is often cited as the greatest environmental disaster caused by mankind.

The Aral Sea once covered an area of 68,000 km^2. It began drying up at a faster rate than its two main feeder rivers could replenish it as soon as water was diverted from the rivers around the sea to **irrigate** cotton fields and arable crops growing in the surrounding region. Traditional fishing communities employing tens of thousands of people, together with 90% of the fish species living in the Aral Sea, were wiped out over a twenty-year period. The drop in the volume of water has also resulted in an increase in the concentration of **pesticides, fertilisers** and herbicides in the soils of the region, which have contaminated drinking water and poisoned the soils that farmers depend on.

'Once a colossal geographic feature – at 26,000 square miles (67,300 square kilometres), it was the fourth largest inland water body on earth in terms of surface area – the Aral shrank to hold just one-tenth of its original volume, becoming a tragic shadow of itself. Its fish died, the most successful fishermen left and those who remained began to starve.'

National Geographic

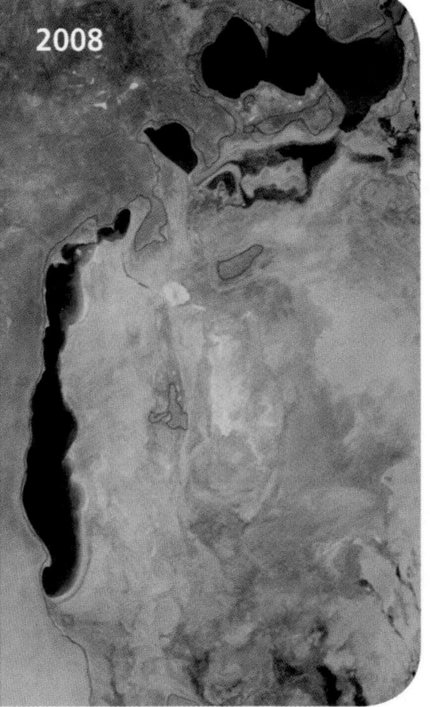

In 2010, a NASA satellite captured a massive dust storm occurring as wind picked up millions of tonnes of soil from the dry lakebed of the Aral Sea and blew it away towards the southeast along the Uzbekistan–Kazakhstan border. The plume of dust is shown as pale beige in the 2010 image.

2010

Since 1994, some progress has been made to combat desertification and restore the Aral Sea. Read the article about the Aral Sea from the UNESCO website.

What has been done to restore part of the north Aral Sea and how are halophyte plants helping to stabilise the soils of the dry seabed?

▶ Consolidating your thinking ◀

How does the information in the table below help to explain why the land degradation in and around the Aral Sea was so devastating?

Indicator	1960	2012
Population of region of Aral Sea (million)	14.1	60.4
Irrigated farming land around the Aral Sea (million hectares)	4.5	8
Water withdrawn from the Aral Sea basin (km³ per year)	60.6	105
Water from tributaries and run off entering the Aral Sea (km³ per year)	55	10.6

Source: EC-IFAS 2013

The United Nations Education, Science and Cultural Organisation (UNESCO) has highlighted the Aral Sea restoration project along with eleven other projects around the world to show what can be done to restore lands which have undergone desertification. Using the template in the Teacher Book (p.45), summarise the key points about each restoration project in Algeria, Gambia, Kenya, Niger, China, India, Chile, Ecuador, Peru, Spain and Italy which are shown on the world map at http://www.unesco.org/mab/doc/ekocd/index_case.html.

4.6 How is the Dartmoor Mires Project helping to combat desertification?

It is not just countries in the dryland areas of the world that are contributing to the fight against desertification. The *Dartmoor Mires Project* is located at South Tavy Head, Dartmoor in one of the wettest places in Northern Europe, with over 2000 mm of precipitation a year, and is more than 4800 km (3000 miles) from Moses and Thandi in Zambia.

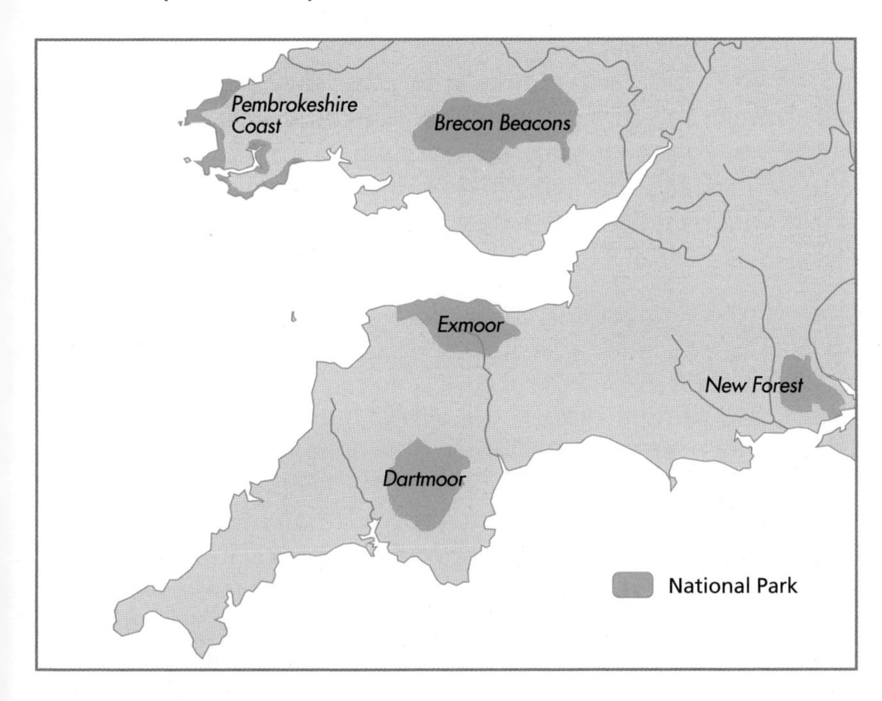

Consolidating your thinking

Working with a partner, examine the maps, graphics and images on this page and watch the short film at http://www.youtube.com/watch?v=fffyO9X2RyA. What do you think the people involved in the Dartmoor Mires Project are doing? How do you think this work could help combat desertification in the dryland areas of the world thousands of kilometres from Dartmoor?

Atmospheric CO_2 concentration from 650,000 years ago to present using ice core proxy data and direct measurements

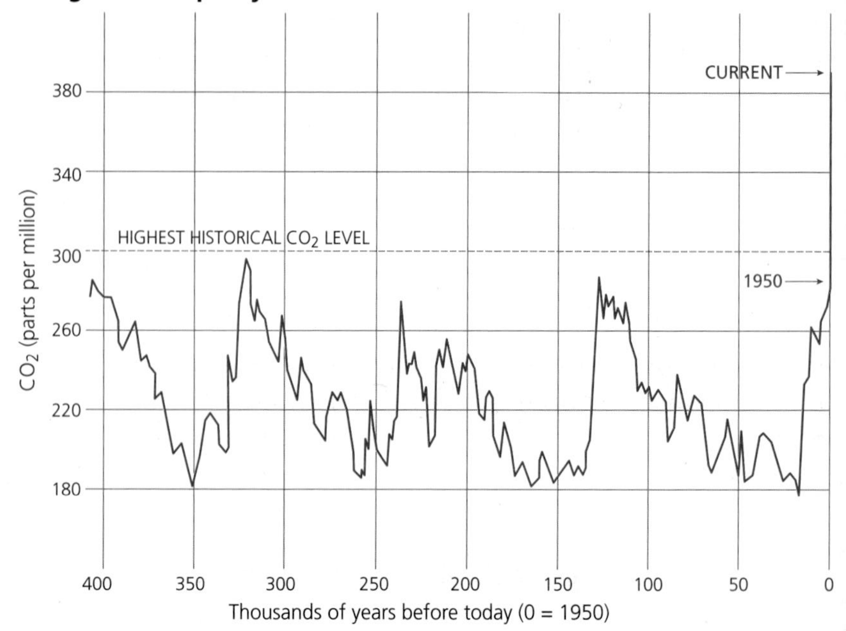

'In Africa carbon dioxide induced climate change and desertification remain inextricably linked. In many African regions the area suitable for agriculture, the length of the growing season and crop yield potential particularly in arid and semi-arid areas are expected to decrease.'

'Climate change and desertification',
World Meteorological Organization (WMO)

In its 2013 annual report, the WMO highlighted the fact that the concentration of carbon dioxide (CO_2) in the earth's atmosphere had reached 142% of the level it was in 1750. In addition, its measurements showed that CO_2 levels increased more between 2012 and 2013 than in any year since 1984. The report identified that this increase was due to reduced CO_2 uptake by the earth's **biosphere** (natural vegetation and soils) as well as to increased emissions (release of CO_2 into the atmosphere). Along with methane and nitrous oxide, CO_2 concentrations contribute to **global warming**, which is in turn linked to changes in rainfall patterns and more frequent extreme weather events, such as prolonged droughts and flash flooding, leading to land degradation and desertification.

The Dartmoor Mires Project is an example of what can be done in countries outside the world's drylands to increase the ability of the planet's biosphere to absorb more carbon from the atmosphere. It has been estimated that ten megatonnes of carbon are stored in the peat soils of Dartmoor: equivalent to an entire year of CO_2 output from UK industry.

If this peat, which has been forming for up to 10,000 years, is left to dry out and erode, then the carbon it stores will be released into the atmosphere, contributing to further **climate change** and desertification in places like Zambia. However, if the blanket bogs are kept in good condition then their unique plant communities will continue to grow, forming layers of new peat and locking away existing carbon, as well as absorbing new CO_2 from the atmosphere.

Protecting and enhancing peatland habitats on Dartmoor

The Dartmoor Mires Project is a five-year pilot project to carry out restoration on areas of internationally important blanket bog, which are threatened by erosion, and then to evaluate the effects of this work.

The project aims to achieve benefits for:

Biodiversity - restoration of a globally important habitat benefiting species such as dunlin.

Water - evaluation of the effects of restoration on the supply and quality of water, in turn benefiting the wider population of the South West.

Carbon Storage - Dartmoor's peatlands store the equivalent of one year's carbon emissions from UK industry. Eroding blanket bog will release this carbon whilst encouraging bog mosses to grow will enable more carbon to be sequestered and stored indefinitely.

Consolidating your thinking

Read the animated cartoon at http://www.unesco.org/mab/doc/ekocd/part1/intro_cartoon.html which is part of the *Environmental Education Kit on Desertification* produced by UNESCO. This kit is aimed at helping younger children around the world to understand the causes of desertification and the measures that can be taken by communities at a local level to prevent it. Make a list of all the things that the children did in their village in Peru which helped to control desertification.

Imagine now that you have been given a brief by Dartmoor National Park to design and produce a similar cartoon or comic strip to explain to children in Peru and Zambia what people far away in the UK are also doing to help with the fight against desertification.

Remember that the cartoon or comic strip will be aimed at 7 to 10-year-old children and must make it clear that you understand the causes and effects of desertification in dryland areas of the world, the kind of things that are already being done in dryland countries to combat it and how the Dartmoor Mires Project is making its own contribution to the battle. You can use copies of the template in the Teacher Book (p.45) to assist with this.

Geographical Enquiry

5 Using someone else's water

How does water consumption create interdependence and conflict?

Increasing population, changing rainfall patterns and higher levels of use are putting areas of the world under high levels of water scarcity. The map on these pages shows regions that are experiencing a scarcity of water for either physical or economic reasons. A **physical scarcity** occurs when more than 75% of river flows are used for human needs (farming, industry and households). An **economic scarcity** might occur when people don't have enough money or infrastructure (wells, pipes, dams and reservoirs) to use the water that is available locally.

5.1 How much water do I use?

There are millions of people who are struggling to access water on a daily basis and the number is growing. This can lead to problems with food supply, **sanitation**, **desertification** and damage to ecosystems. Many rivers flow across national borders, so conflict over access to freshwater in the future is inevitable.

People in the United Kingdom use an average of 160 litres of water each per day. This is more than the average use per person of Burkina Faso, Niger, Angola, Cambodia, Ethiopia, Haiti, Rwanda, Uganda and Mozambique added together. The average person in the USA uses an astonishing 575 litres each per day. Thinking about and reducing our own water use is important because even within countries with plentiful water supplies there can often be regions that suffer shortages. For example, Southeast England has less water per head than Sudan in North Africa.

In addition to this, there are financial and environmental costs (pollution and CO_2 emissions, for example) involved in collecting, treating and pumping water to homes, as well as potential damage to ecosystems by over-abstraction of water. Thames Water Utilities Ltd has even built a desalination plant to use in times of high demand in London. These plants use huge amounts of energy.

Global water scarcity

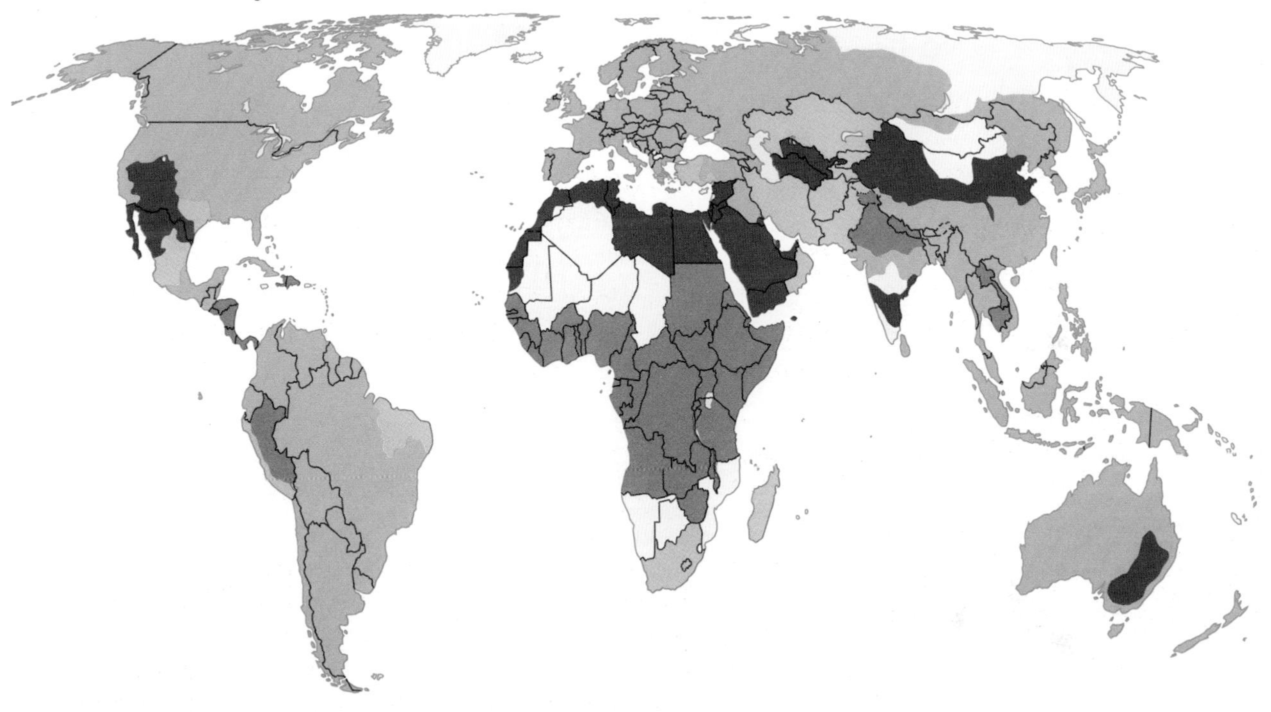

Water scarcity
- Extreme physical water shortage
- Physical or approaching physical water shortage
- Economic water shortage
- Little or no water scarcity
- No data

A desalination plant in Hamburg, Germany

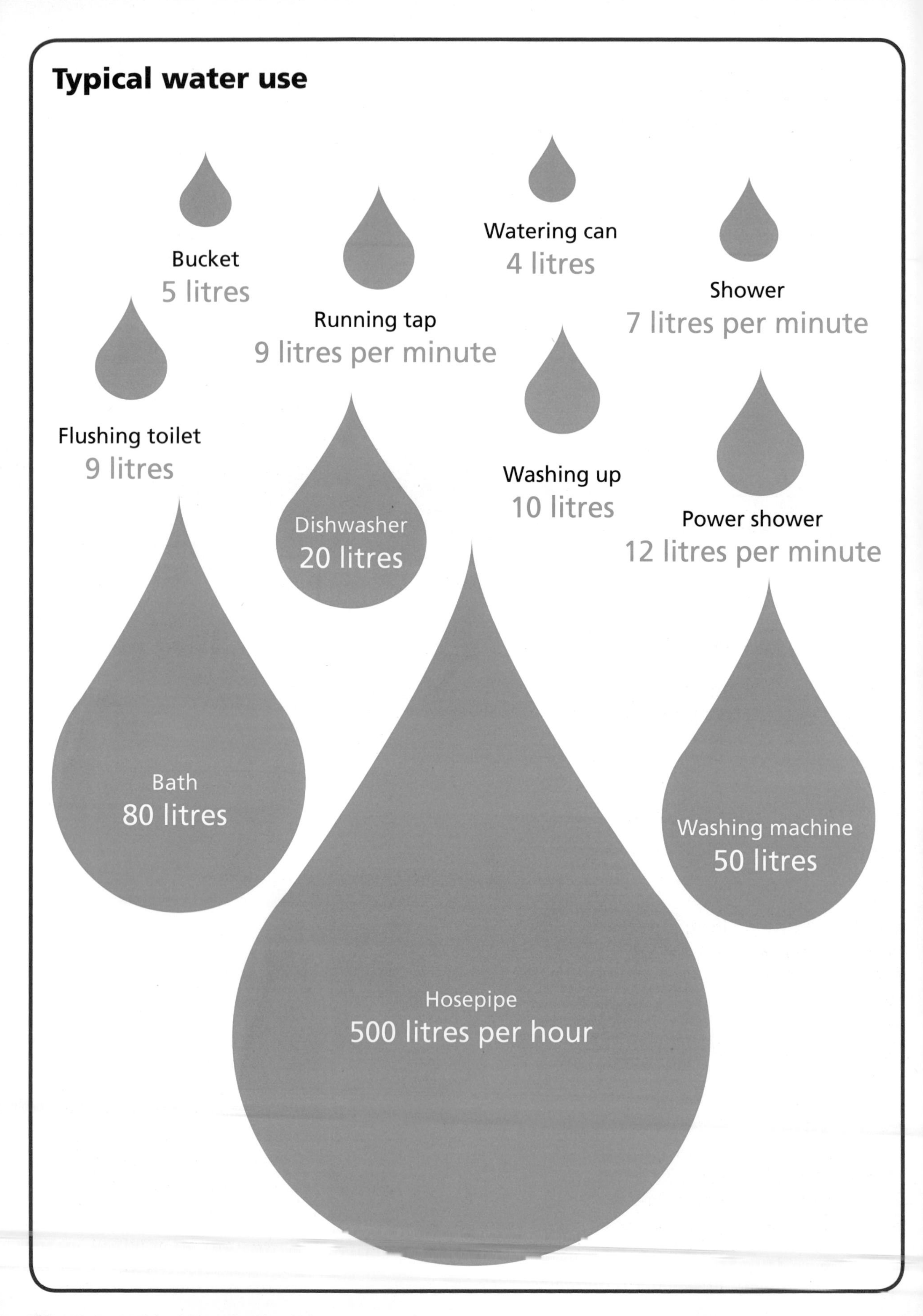
Typical water use
Bucket
5 litres
Running tap
9 litres per minute
Watering can
4 litres
Shower
7 litres per minute
Flushing toilet
9 litres
Washing up
10 litres
Dishwasher
20 litres
Power shower
12 litres per minute
Bath
80 litres
Washing machine
50 litres
Hosepipe
500 litres per hour

Average water consumption

per person per day (litres)

Bangladesh

Kenya

Ghana

Nigeria

Burkina Faso

Niger

Angola

Cambodia

Ethiopia

Haiti

Rwanda

Uganda

Mozambique

▶ Consolidating your thinking ◀

Make a list of all the things you use water for on a daily basis. Can you survive without water? Look carefully at the map on page 73. Use the copy of the map from the Teacher Book (p.49) to describe the location of areas that are suffering from water scarcity. Can you suggest possible reasons why this is happening?

You should then add up how much water you have used so far today using the typical water **consumption** figures. Compare this with others in your class. Use the bar graph template on page 48 of the Teacher Book to show your own consumption and compare this to the consumption figures for other countries. Finally, you should design a poster to encourage others at school to reduce their water consumption. Ensure that you include practical tips and explain why it is important.

Extending your enquiry

You could complete a week's water diary for your home. Total up the consumption and work out the average daily consumption per person. Find out the utility company providing water to your school and look online to discover whether they make any suggestions about how consumers can conserve water.

Managing consumption is an important part of acting sustainably. You could find out how to save more water by playing the game at this link:

www.thewaterfamily.co.uk

One packet of crisps requires 185 litres of water to grow the potatoes and complete the production process. It takes about 140 litres to grow one cup of coffee, about 11,000 litres to produce a pair of jeans and about 400,000 litres to build a car.

It is important to consider how one cup of coffee uses 140 litres of water, so you should consolidate your thinking by using the example of beef in Figure B and by using this link:

http://www.angelamorelli.com/water/

Some useful web links:

https://itunes.apple.com/app/virtual-water/id369876250?mt=8

http://www.waterfootprint.org/?page=files/home

http://environment.nationalgeographic.com/environment/freshwater/embedded-water/

http://www.waterwise.org.uk/pages/embedded-water.html

http://www.virtual-water.org/index.php?option=com_wrapper&Itemid=8

Distribution of Global Water Resources

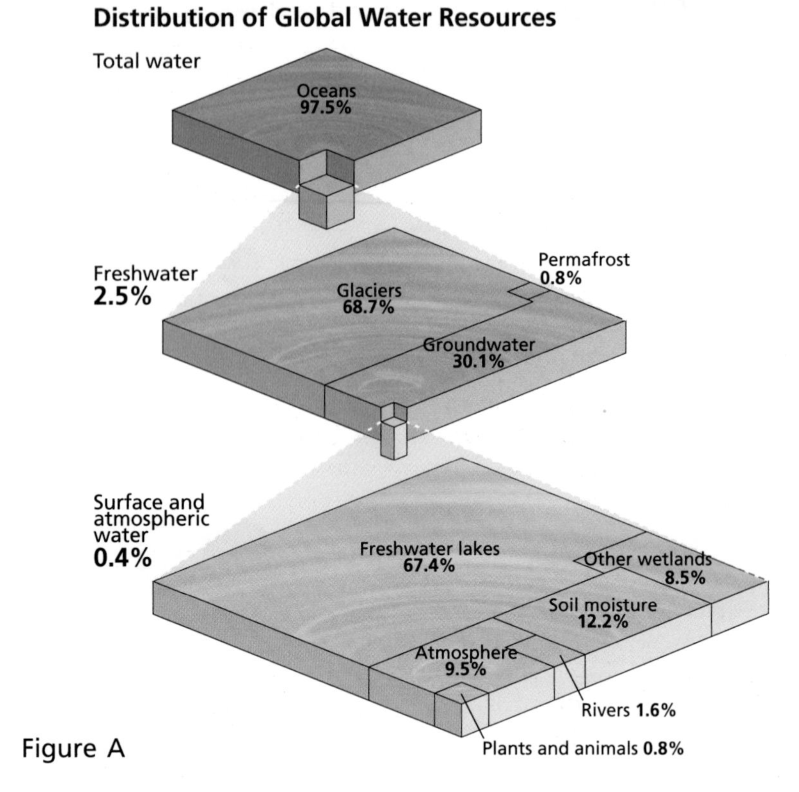

Figure A

Examine Figure A closely. We live on a blue planet with two-thirds of the surface covered in water. Despite the abundance of water, only 1% of it is available for human consumption. This is shared between the atmosphere, soil, plants, animals, rivers and freshwater lakes and is not evenly distributed. The rest is locked up in the oceans, ice caps and under the ground. A growing global population and rising levels of development are placing increasing pressure on water as a global resource.

On page 72 you learned that the average person in the UK uses 160 litres of water per day for washing, drinking and cooking. However, a lot of the UK's food and clothes are now grown or made in other countries. When taking into account the water used in the production of food and clothes, the average water use in the UK rockets to an astonishing 3400 litres per person per day! That is the equivalent to over one million one-litre bottles of water each per year – with most of these coming in from other countries that are struggling with water scarcity! The amount of water used to produce a product is referred to as embedded or **virtual water**.

Some scientists are forecasting that when the world's population rises beyond eight billion in the next ten to fifteen years, the global demand for food and energy will jump by 50%, with the need for fresh water rising by 30%. Much of this growth will occur in less economically advanced areas of the world which are already using significant amounts of their water to grow food and manufacture products for consumption in more advanced economies such as the UK and USA. This may lead to products becoming too expensive or simply unavailable in the future.

Using water in **water-stressed** Kenya to grow our flowers and beans might be unsustainable, but does provide the country with valuable money from **exports**. Consumers and farmers are **interdependent**. For example, a drought in Kenya will cause shortages of particular crops in the UK whilst a decline in popularity for a particular product in the UK will force changes on farmers. Managing this problem is complex and climate change means that our use of virtual water and its impact in water-stressed countries will continue to grow.

Consolidating your thinking

Now you are going to design your own enquiry into virtual water. You can come up with your own questions or use these examples to help your thinking:

- What is virtual water and why is it a complex issue?
- Why should we consider the virtual water in the products we consume?
- Who is affected? How would Kenyan flower farmers feel about exporting products to the UK? Consider that they might have unreliable and unclean water supplies for their families.

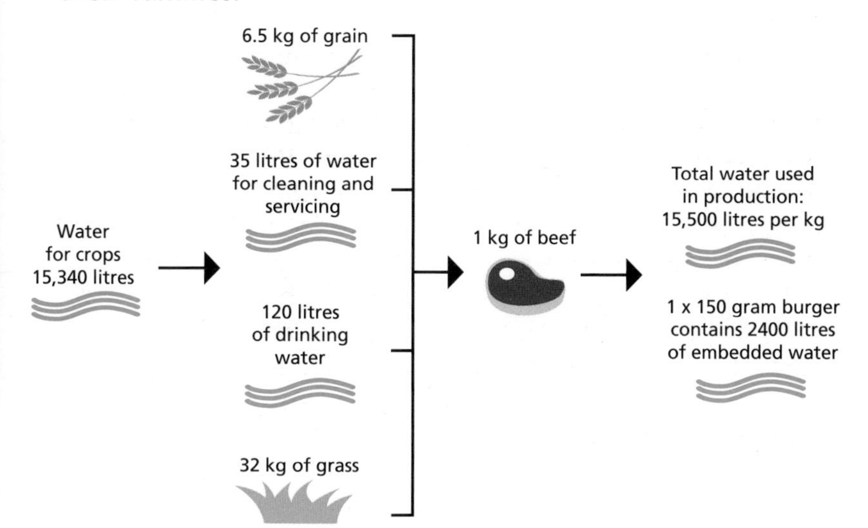

Figure B: *Diagram showing how many litres of water are consumed in the production of 1 kg of beef*

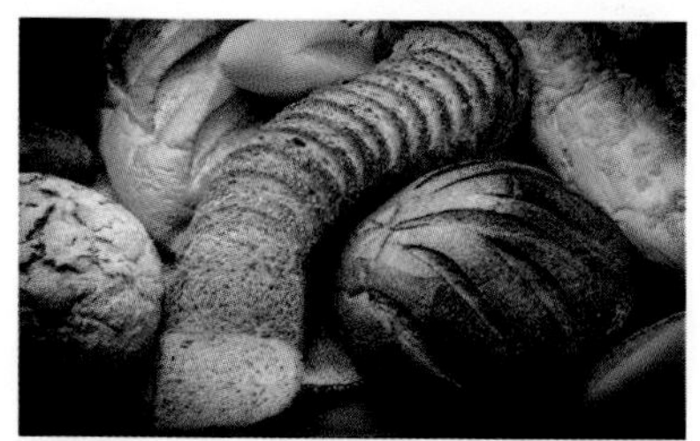

Bread
1 slice = 40 litres of water

Beef
1kg = 15,500 litres of water

Coffee
1 cup = 140 litres of water

Leather shoes
1 pair = 8000 litres of water

Milk
1 glass = 200 litres of water

Burgers
1 = 2400 litres of water

5.3 Why is Marisa catching fog?

Marisa Sanchez lives in Lima, the capital city of Peru in South America. She lives with her two children in a slum, known as a **barriada**, on a steep hill overlooking the city. Lima is the driest and highest capital city in the world – it extends from the **coastal plain** up into valleys and mountain slopes as high as 1550 metres above sea level. Lima has a population of over eight million people with one-third of these living in barriadas.

Marisa doesn't have running water, so she has to rely on expensive water deliveries from a truck that has to navigate the steep and unstable roads constructed by the residents of the barriada. The residents in these areas pay significantly more for water than anyone else in the city, despite being the poorest residents. Marisa has an outside toilet which drains into a pit she has dug herself. *Why* do Marisa and millions like her in Lima lack access to water? Is it simply a lack of rain, money and poor infrastructure or something else?

Use the images here to consider how a lack of water would impact Marisa and her children. Why do you think she lacks access to water?

PERU FACTS

Capital and largest city: Lima

Official languages: Spanish, Quechua and Aymara

Area: 1,285,216 km^2

Population: 30,814,175

GDP: US$11,988 per capita

HDI: 0.737

Currency: Nuevo sol

Birth rate: 20 per 1000

Death rate: 5.3 per 1000

Life expectancy: 74.5 years

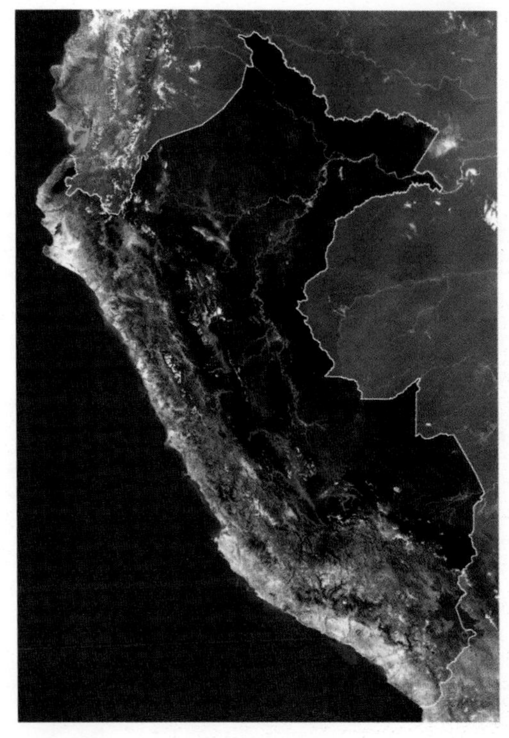

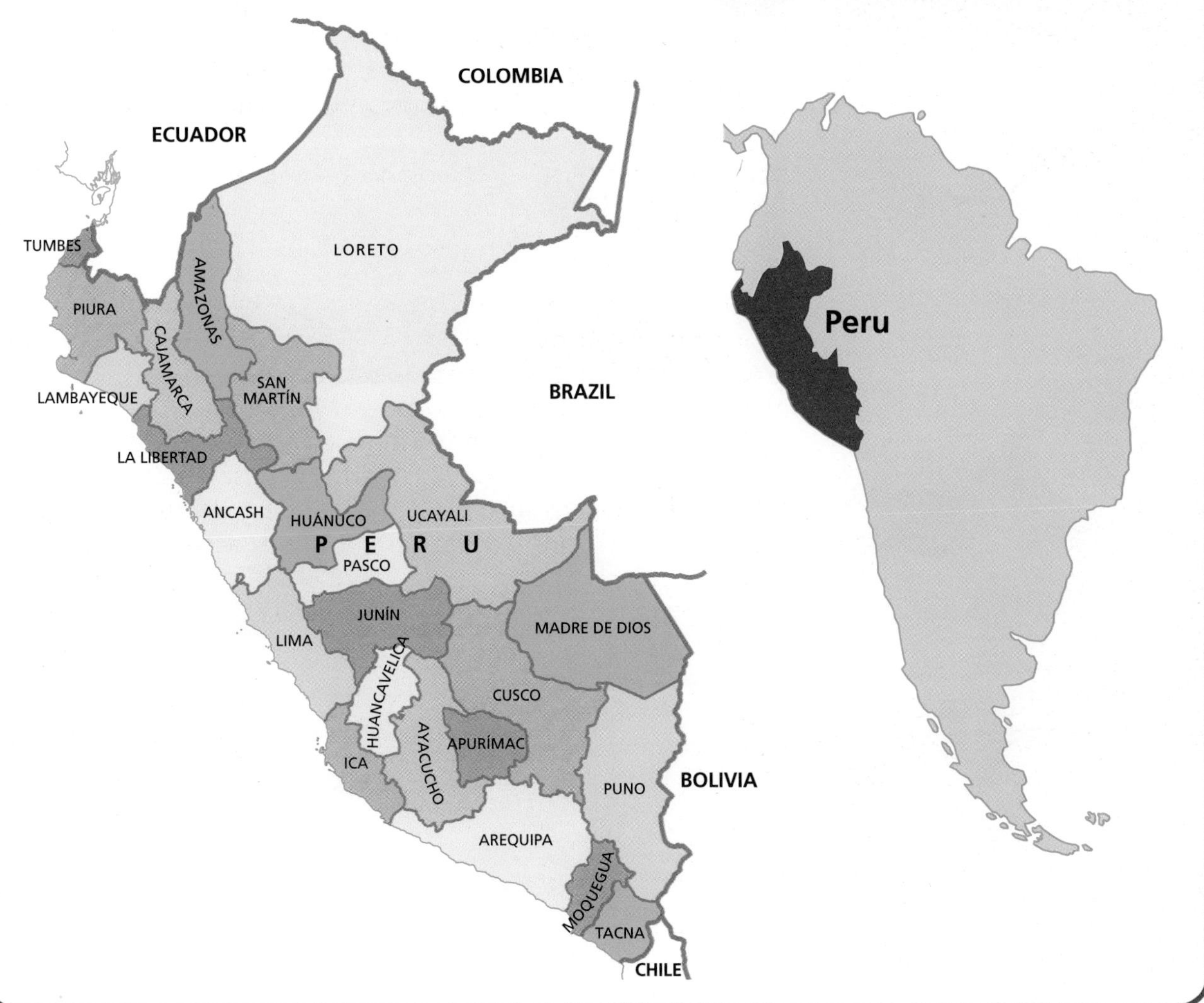

- What is the connection between these maps and images? Can you see any patterns?
- Using the **population density** and **precipitation** maps, what proportion of Peru's population live in areas with low rainfall?
- Suggest reasons why this might explain the lack of water in Lima.

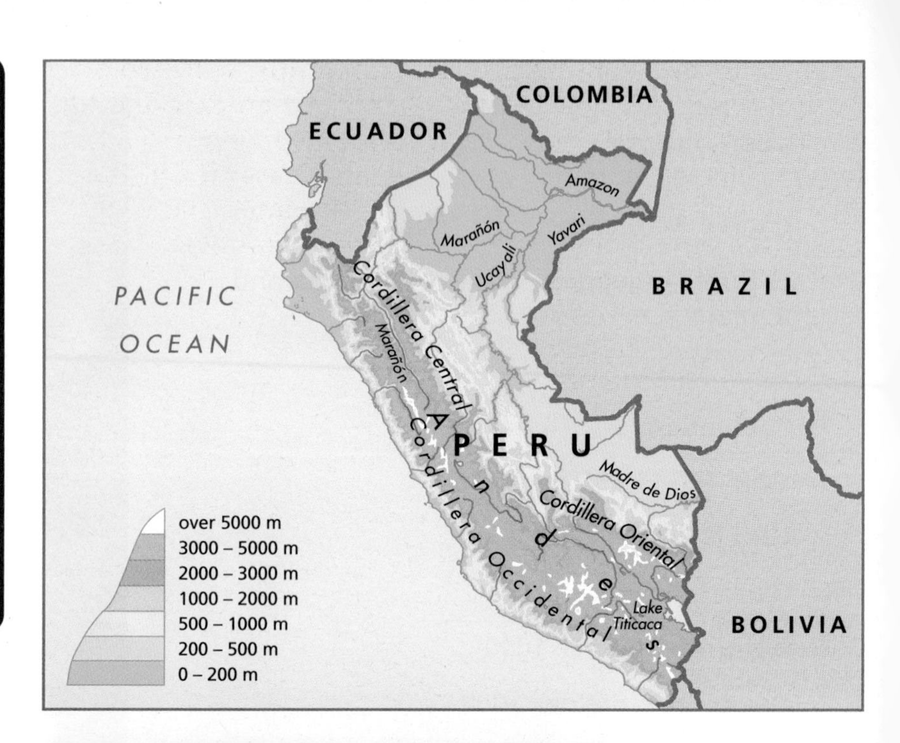

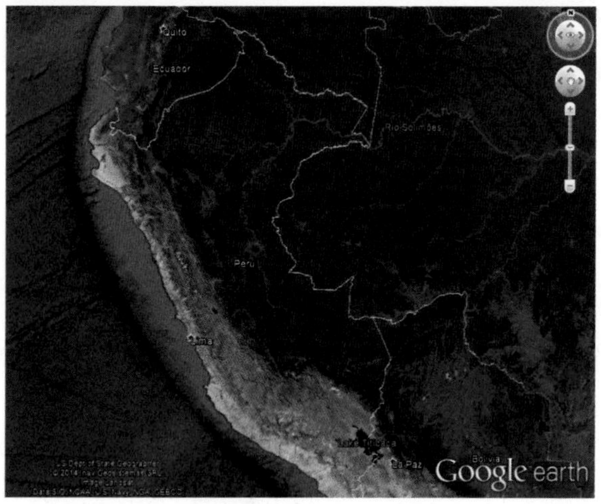

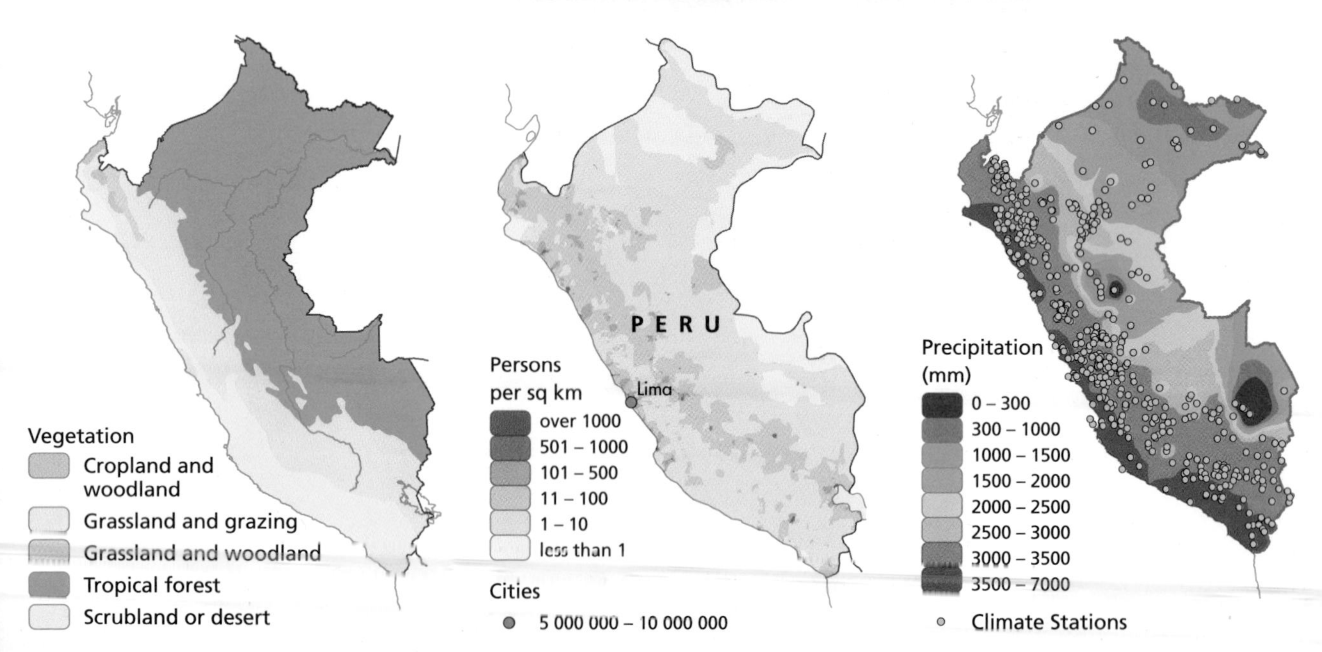

Lima has an **arid subtropical climate,** but due to the cold Pacific waters, it is much cooler than might be expected for a subtropical desert. Humidity is always high, which leads to fog on summer mornings when the land is warm and the air cool, and persistent low cloud in the winter. The cool, stable, moist air from the Pacific Ocean warms over the land and rarely gives rain.

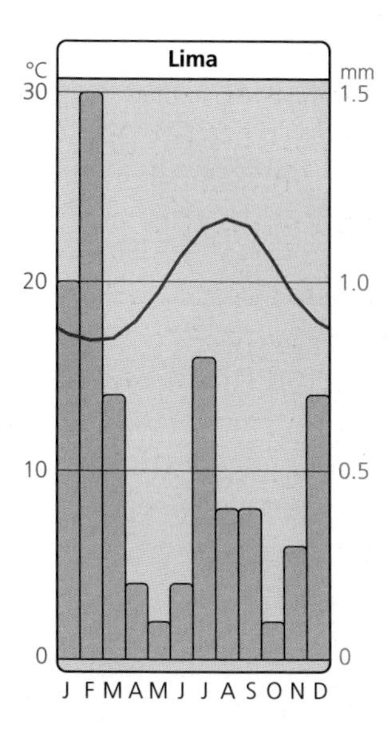

Figure A: *Low cloud on a winter's day in Lima*

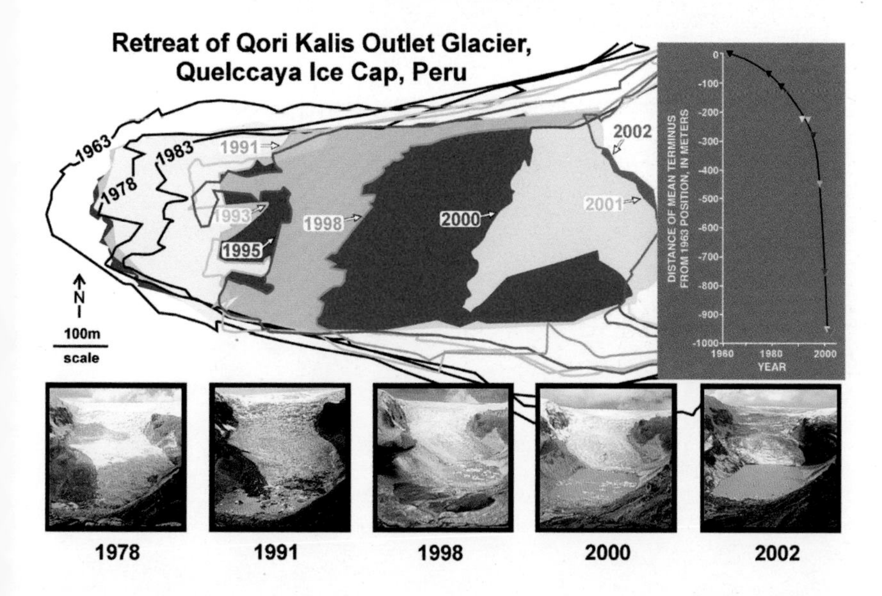

Many of the rivers in Peru are fed by **glaciers** from **ice caps** high in the Andes.

- Describe what has happened to the Qori Kalis glacier since 1978. Why is this happening?

- How might this impact the situation in Lima?

Write your answers on a copy of page 55 of the Teacher Book. You could print off images or draw diagrams to illustrate your answers.

The Rimac River has a critical function in Peru. It provides farmers in the mountains with water for drinking and crops before being used to support vital mining industries (which can pollute the water). It is used to generate electricity and then finally as drinking water in Lima.

- Look carefully at Figure A, which shows the main intake for Lima's drinking water. What is happening on the left of the image?
- What does this tell us?

Figure A

Water droplets from fog have condensed on this tree

When considering all the geographical evidence together, can you explain why Lima suffers from a scarcity of water? Why is Marisa catching fog?

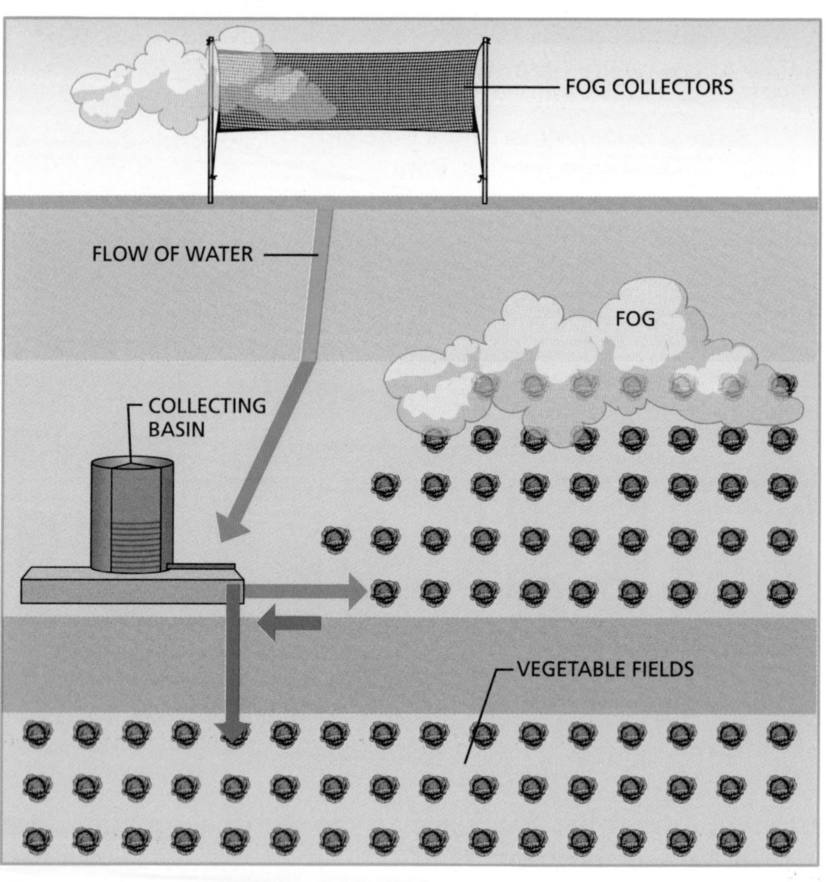

*The diagram above shows how giant nets can be used to create artificial **condensation** surfaces. The water droplets condense on the nets and then fall to the bottom to be collected for drinking or channelled to crops.*

Consolidating your thinking

You need to draft a piece of explanatory writing to bring all of this information together and help you answer Question 5.3: '*Why is Marisa catching fog?*' Use the information you have gathered from the last few pages, plus additional research of your own, to demonstrate that you understand the geographical reasons which explain why Marisa and her community have to catch the fog that often blankets Lima. Your explanatory narrative needs to have the following structure:

- **A title:** Why is Marisa catching fog?

- **An introductory paragraph** to set the scene and context – in this case providing background information on Marisa, where she lives and the life she leads. You will need to consider whether you will use maps and images to help set the context.

- **A second paragraph**, which begins with a **topic sentence** (this introduces the reader to what the paragraph is going to be about). In this paragraph you will discuss the physical geography of Peru and link this with appropriate maps and images.

- **A third paragraph** (the focus of which will again be introduced via a topic sentence), which explains the impact of the physical geography of Peru on human activity (where and how people are able to live, for example) and how the changing environment is impacting water supply in Lima. You must use connectives such as 'since', 'because', 'so', 'as', 'therefore' and then 'this leads to', 'which causes', 'this means', 'as a result of', 'due to the fact that', etc.

- **A concluding paragraph**, which is a summary of the main points and answers the question. Once again, look to apply appropriate connectives such as 'in conclusion', 'in summary', 'to sum up', 'overall', 'on the whole', 'in short', 'in brief', 'to conclude', 'so to round off', etc.

During the drafting stage of this piece of explanatory writing, ensure that you share your work with a peer and ask them to assess your draft using the peer assessment sheet in the Teacher Book (p.53). Give them feedback on what went well and make sure you include suggestions for improvement using the heading 'Even better if...'

5.4 How does water consumption create interdependence and conflict in Peru?

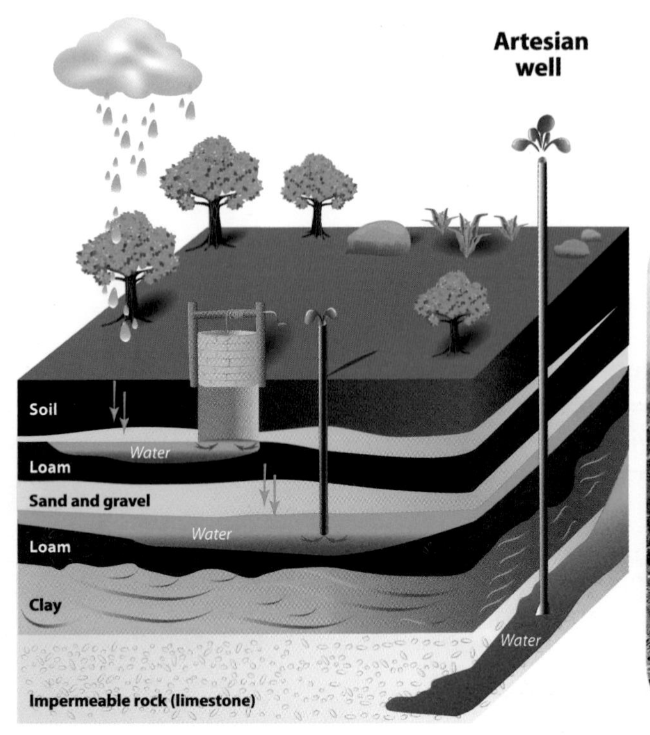

This lush green valley floor is in the Ica Valley, southeast of Lima. Despite being in the arid coastal plain, this area is vital for growing crops – not for Peruvians, but for export to developed nations in Europe and North America.

Underground water stores, known as **aquifers**, are fed by melting ice and rivers. Water passing over permeable ground will **infiltrate** the soil and then **percolate** into the underlying **porous rock**. The top of the underground area that is saturated or full of water is known as the **water table**. Using boreholes, it is possible to extract this water and to irrigate dry land, which can be used for growing crops. However, this can reduce water availability downstream and cause pollution from pesticides and fertilisers.

Due to the **abstraction** of water here, which has been financed by the World Bank, Peru has been able to grow asparagus all year round. This once seasonal vegetable is in high demand in rich countries and Peru has become the largest exporter of asparagus in the world. Consider the following:

- Peruvian asparagus export will make an estimated US$570 million in 2014.

- The industry employs over 10,000 people.

- Many of the farms are owned by large corporations. These large-scale export farms have been allowed to buy up the rights to water supplies, leaving small farms reliant on wells that are drying up.

- The USA is the largest importer of asparagus, with over 174 million lbs imported annually.

- Asparagus is a thirsty vegetable; the USA water footprint resulting from importing asparagus is an incredible 93 million m^3 of Peruvian water a year – enough to fill 37,000 Olympic swimming pools.

- The asparagus industry abstracts over 317 million m^3 per year from the Ica Valley aquifer.

- This intensive **irrigation** does not use the latest technology, so can be wasteful. The aquifer has fallen by as much as 8 m in some places.

- Villagers surrounding the Ica Valley have reported difficulties in accessing drinking water and water to feed their own crop.

- Farmers are noticing the loss of water, so are beginning to change crops – blueberries, grapes and pomegranates are all being increasingly grown for export.

▶ Consolidating your thinking ◀

Intensive agriculture is clearly bringing many benefits to Peru, but at what cost? As consumers, when we buy asparagus from Peru we become interconnected with the farmers and the people affected by the depletion of their water resources. You will need to work with a partner to produce an infographic on Peru, covering agriculture, exports, interdependence and water conflicts. It is important that your work addresses Question 5.4: '*How does water consumption create interdependence and conflict in Peru?*'

You should consider the context of agriculture in a country suffering from both a physical and economic scarcity of water, the costs and benefits of export-led intensive farming, the interdependence between farmers, villagers and consumers, and whether you feel this is a sustainable industry. If not, you should think about suggestions for making it more sustainable.

The Olmos Irrigation Project

A dam high in the Andes

A water transfer tunnel

Olmos irrigation scheme
International boundary
Road
New access road
Railway
Lima tunnel
Port
Airport
Capital city

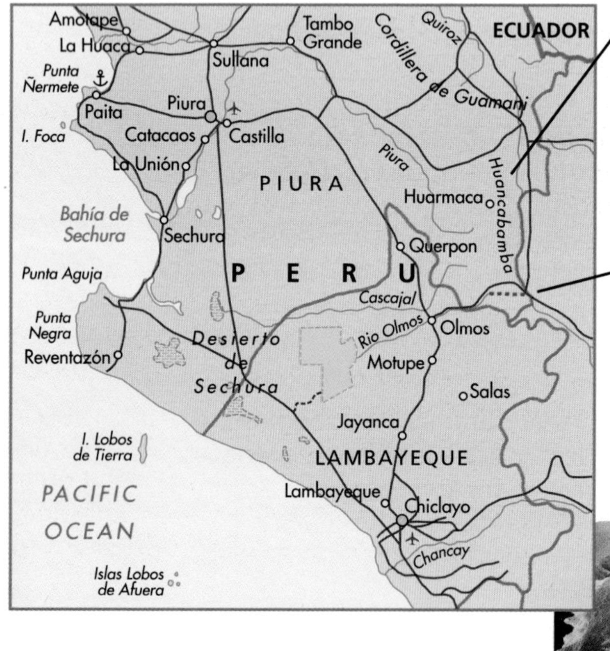

The Rio Huancabamba drains east into the Amazon basin and the Atlantic

Location of 20 km tunnel and power stations. The Limón Dam and tunnel will divert some of the flow into the Rio Olmos

The landscape will soon have irrigation canals full of water like this one in Arizona, USA

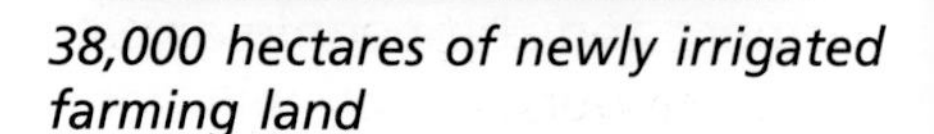

38,000 hectares of newly irrigated farming land

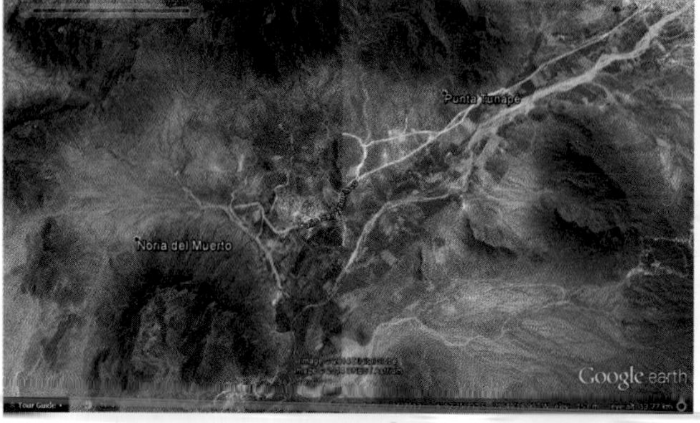

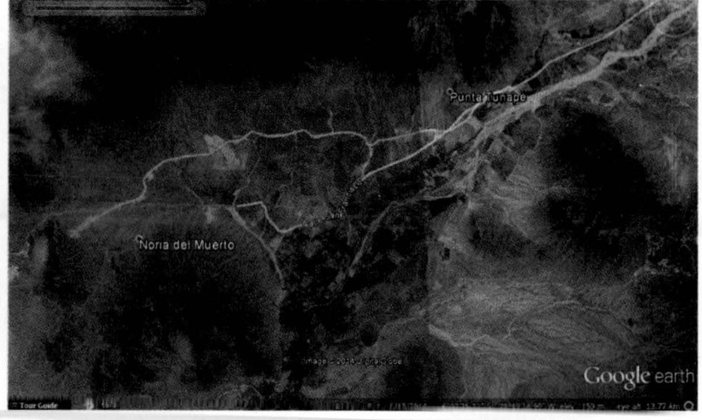

These two satellite images, from different satellites, show the area to be irrigated in July 2013 (L) and June 2014 (R). Can you identify any evidence of the irrigation project?

Extending your enquiry

The Peruvian government is acutely aware of its vulnerable position in terms of water. Intensive agriculture, using aquifers fed by rivers that are heavily used and glaciers that are retreating, is a huge growth area for the economy and provides jobs and foreign income. With predictions of rising temperatures and changing rainfall patterns, the government is beginning to take action. The US$500 million **Olmos Irrigation Project** is a huge engineering scheme to transfer water from one drainage basin to another through a 20 km sequence of massive tunnels and rivers.

This is the first irrigation scheme in Peru to take water destined for the Amazon and move it across the Andes. The water will be used to create nearly 40,000 hectares of irrigated agricultural land in Northern Peru, as well as to generate clean **hydroelectric power**. Supporters say there will be many direct and indirect benefits to people and the economy. Critics point to the fact that it only benefits big farms used for export.

Look carefully at the scheme details and the location of the irrigation project. Using these, as well as your knowledge and understanding of Peru, virtual water and water consumption, consider the following questions:

- What will this **water transfer scheme** be used for?
- Why is it being located here?
- Will this benefit Marisa Sanchez and her children?

Use the template on page 54 of the Teacher Book to record your answers.

You can investigate further here:

http://odebrecht.com/en/olmos-irrigation-project-odebrecht-begins-water-transfer-northern-peru

http://uk.reuters.com/article/2013/04/04/us-peru-water-idUSBRE9330QT20130404

http://www.dailymaverick.co.za/article/2013-04-05-peru-bores-through-andes-to-water-desert-after-century-of-dreams/#.VEd0ZCKUeSo

http://www.asce.org/cemagazine/Article.aspx?id=23622325766#.VEdx9yKUeSo

Geographical Enquiry

6 Biodiversity under threat

Can economic development on Borneo be sustainable?

The 2014 **World Wide Fund for Nature (WWF)** Living Planet Index reported that there has been more than a 50% decline in animal population between 1970 and 2010. The report clearly states that 'the biggest recorded threat to **biodiversity** globally comes from the combined impacts of habitat loss and degradation, driven by unsustainable human **consumption**.'

Some facts to consider:

- Sea ice extent in the Arctic is shrinking at a rate of 3% every ten years.
- Since 1980, humans have cut down over 225 million hectares of forest (nine times the size of the UK!).
- An average person in Mozambique (Africa) uses four litres of water per day. In the USA this figure is an astonishing 575 litres.
- The amount of carbon dioxide (CO_2) in the atmosphere is higher now than at any point for the last 650,000 years.
- Cutting down one hectare of rainforest releases 500 tonnes of CO_2 into the atmosphere – the same as driving a car around the equator twenty-eight times!

6.1 What does 'sustainable' mean?

The island of **Borneo** in Asia is one place where human activity is having a profound impact on biodiversity. During this enquiry you will learn about Borneo and how it is being affected by our consumption of resources. However, analysing, evaluating and concluding whether this is sustainable is a complex process, so this enquiry first examines the concept of sustainability itself.

Humans are using ever more of the earth's resources in ways that will affect the quality of life of some people, both today and in the future. **Sustainability** is about allowing us to meet our needs today without affecting the chances of people in the future to meet their own needs. Human actions can be seen as **sustainable** or **unsustainable**.

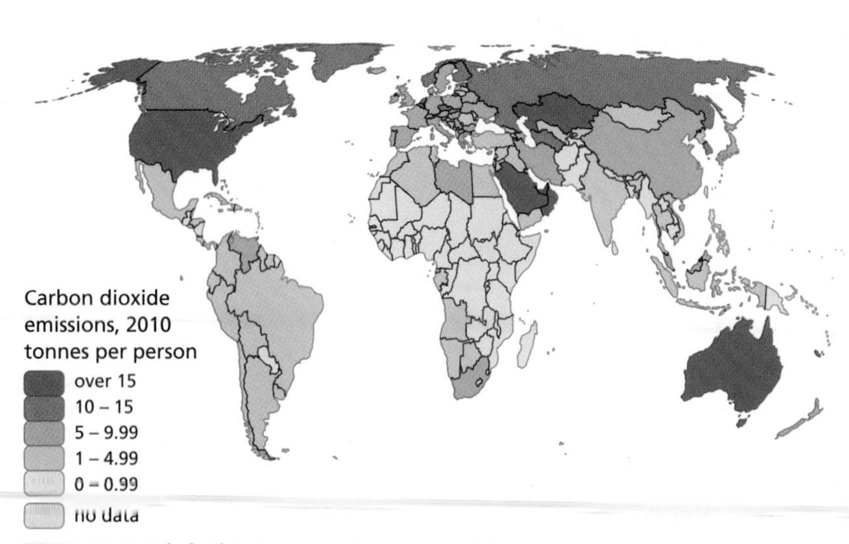

Figure A: *Global CO_2 emissions in 2000*

As levels of wealth grow around the world, global demand for products increases. This rise in consumption is placing increasing pressure on the earth's resources and is having an ever more damaging impact on the natural environment. Acting in a more sustainable way allows us to reduce this impact. Humans are using a lot of fossil fuels. These release CO_2 into the atmosphere and contribute to the greenhouse effect. Figure A shows countries of the world according to their CO_2 footprint; bigger countries have inflated footprints. Figure B shows how many planets we would need to continue to live the average lifestyle in different countries. If we all consumed like the USA, we would need 5.3 planets. **This is unsustainable**.

Global footprint

USA	UK	South Africa	China	World average
5.0 Planets	**3.4 Planets**	**1.5 Planets**	**1.0 Planets**	**1.4 Planets**

Figure B: *Eco footprints – how many planets do we need?*

Consolidating your thinking

Firstly, make a mind map to show how humans impact the planet. Swap your results with someone else in the class to compare and to build up even more ideas. You should then examine the following human activities and decide whether they are sustainable or unsustainable. You must explain why you think this.

- Using cars to make short journeys rather than walking.
- Leaving the lights on all day.
- Having a short shower instead of a bath.
- Burying all our waste in a landfill site.
- Using wind power to generate all our electricity.

You should attempt to come up with your own examples of sustainable or unsustainable human activities. Record your answers on page 58 of the Teacher Book. Finally, look again at Figure A. Can you describe and explain any pattern that you see?

The more resources we consume, the more waste we produce and the more space we need. Ever-expanding consumption and waste on one un-expanding Earth is unsustainable. Deciding whether an activity is sustainable or unsustainable is complicated because human activities affect different groups of people and places in different ways. However, it is possible to use *geographical skills* to help us in our *decision-making*.

Figure A: *A three-legged stool*

For a human activity to be considered sustainable there has to be a *balance* between the effects on people, the economy and the environment. A good method of judging an activity is to think of it as a three-legged stool (Figure A). If we take one of the legs of the stool away, then it will fall over and the activity will not be sustainable. For example, building a new open cast coal mine:

- Provides jobs for people (social benefit).
- May help the country make money through trade and reducing the need to import coal from other countries (economic benefit).
- Could damage the local environment and, when the coal is burned, lead to an increase in global levels of CO_2 (environmental disadvantage).

Is this a sustainable human activity? Why is it difficult to decide?

Who decides?
These are political questions about decision making, conflict, costs and benefits.

Natural
These are questions about the environment – natural systems: water, air, plants and animals and their relationship. Also, questions about the human environment and the impact on the natural environment.

Social
These are questions about people, society and culture. You can ask questions about race, gender and age.

Economic
These are questions about finance, economies, aid, trade, supply and demand.

Figure B: *The development compass rose*

Consolidating your thinking

We can use images to judge how sustainable an activity is by thinking of *geographical questions*. By using the 'Five Ws' (who, what, where, when and why) and the development compass rose (Figure B), we can think of different questions about a range of factors, including the natural environment, money, people, power and decisions.

Figure C: *Hoover Dam, Nevada USA*

Examine Figure C closely. Using the learning activity from the Teacher Book (p.59), you will need to decide if this is a sustainable or unsustainable activity. Work with a partner and explain what you think is happening in the photo. You should write a description of what you can see and identify all the things that might be sustainable or unsustainable about the image. Give a reason for each one. You will then need to write down at least four geographical questions about the image using the development compass. Finally, you should write a paragraph stating whether you think it is an image of something sustainable or unsustainable. Explain how you reached your decision.

You could then research answers to your geographical questions and present these to the class. To develop your skills further, repeat this exercise using another copy of the learning activity and images from elsewhere in this book, or you could choose your own images. You should then present your image to the class and use the peer assessment sheet from the Teacher Book (p.60) to assess each other's understanding of the images.

6.2 What is special about the island of Borneo?

Driven by economic growth and global demand for goods produced there, the island of Borneo is one place that has seen a significant environmental impact as a result of human activity. There are many places in the world that are impacted by human activity, but Borneo is a biodiversity hotspot – home to pristine **tropical forests** and countless unique species of plants, animals and insects. This astonishing biodiversity is under threat from a massive level of **deforestation** as the people of Borneo strive to make a living and to participate in the global economy.

Consolidating your thinking

As geographers investigating a place and issue, it is vital to gain a sense of what the place is like through the use of maps, graphs, images and statistics. You will need to research Borneo using the information and images on the following pages, plus your own research, to produce an A3 poster around the theme of '*What is special about the island of Borneo?*' If you use digital technology, you could use Google Docs to work on this collaboratively with a partner. There is a blank map of Borneo on page 61 of the Teacher Book which you can use illustrate your poster.

Borneo Factfile

On average, **three** new species are discovered each month on Borneo.

Medical researchers are discovering many uses for plants.

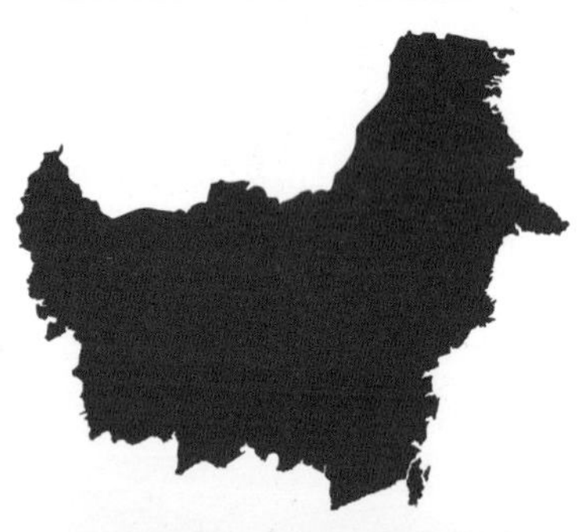

Location: Southeast Asia (south of the South China Sea)

Borneo is shared by 3 countries – Indonesia (Kalimantan) 73%, Malaysia (Sabah and Sarawak) 26% and Brunei 1%

Area: 743,330 km^2 – the world's third-largest island

Population: 19,804,064 (as of 2010)

Population density: 21.52 people per km^2

Highest point: Mount Kinabalu – 4095 m

Longest river: Kapuas in West Kalimantan – 1143 km

Climate: Equatorial

The Borneo Rainforest is **140 million** years old.

There are over **15,000** species of plants and **5000** of these can only be found in Borneo.

3000 species of trees, **221** species of mammals, **420** species of birds and over **440** different species of freshwater fish are found on the island.

There are carnivorous plants, the world's largest flower and trees that reach up to **60 m** in height.

3000 – 5000 m
2000 – 3000 m
1000 – 2000 m
500 – 1000 m
200 – 500 m
0 – 200 m
International boundary
Road
Railway
Airport
Capital city
MYANMAR
Nay Pyi Taw
THAILAND
Bangkok
LAOS
Vientiane
VIETNAM
Ha Nôi
CAMBODIA
Phnom Penh
CHINA
MALAYSIA
Kuala Lumpur
Putrajaya
Singapore
SINGAPORE
Bandar Seri Begawan
BRUNEI
PHILIPPINES
Manila
INDONESIA
Jakarta
Dili
EAST TIMOR
PAPUA NEW GUINEA
AUSTRALIA
South China Sea
PACIFIC OCEAN
INDIAN OCEAN
Andaman Sea
Gulf of Thailand
Sulu Sea
Celebes Sea
Java Sea
Flores Sea
Banda Sea
Molucca Sea
Ceram Sea
Arafura Sea
Timor Sea
Savu Sea
Strait of Malacca
Macassar Strait
Luzon Strait
Gulf of Tongking
Borneo
Sumatra
Jawa
Celebes
New Guinea
Luzon
Mindanao
Halmahera
Hainan
SABAH
SARAWAK
KALIMANTAN
Peninsular Malaysia

There are seven distinct **ecoregions** in Borneo:

- lowland rainforest
- peat swamp forest
- kerengas heath forest
- freshwater swamp forest
- Sunda Shelf mangroves
- montane rainforest
- alpine meadow

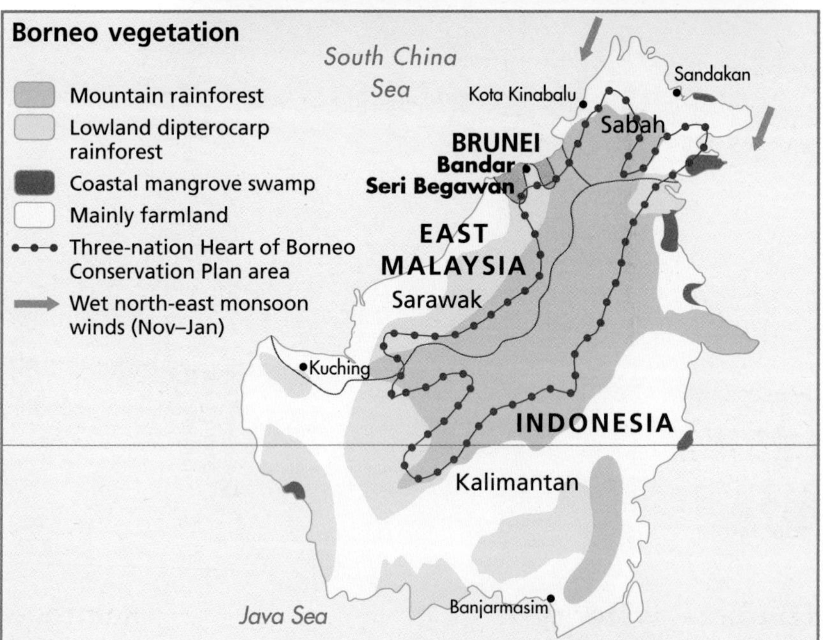

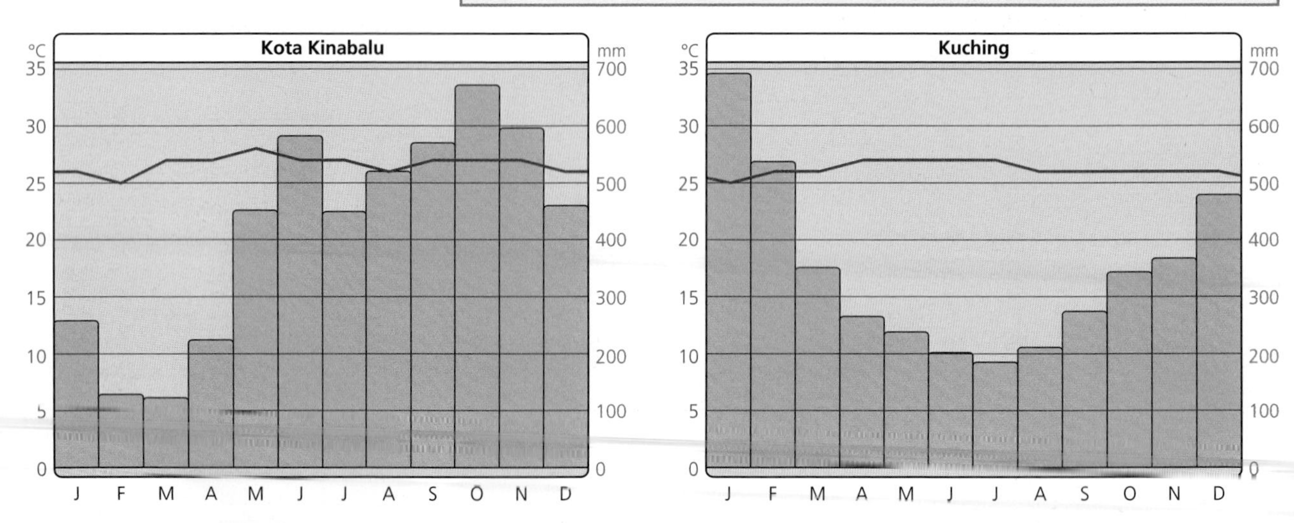

Layers of rainforest vegetation

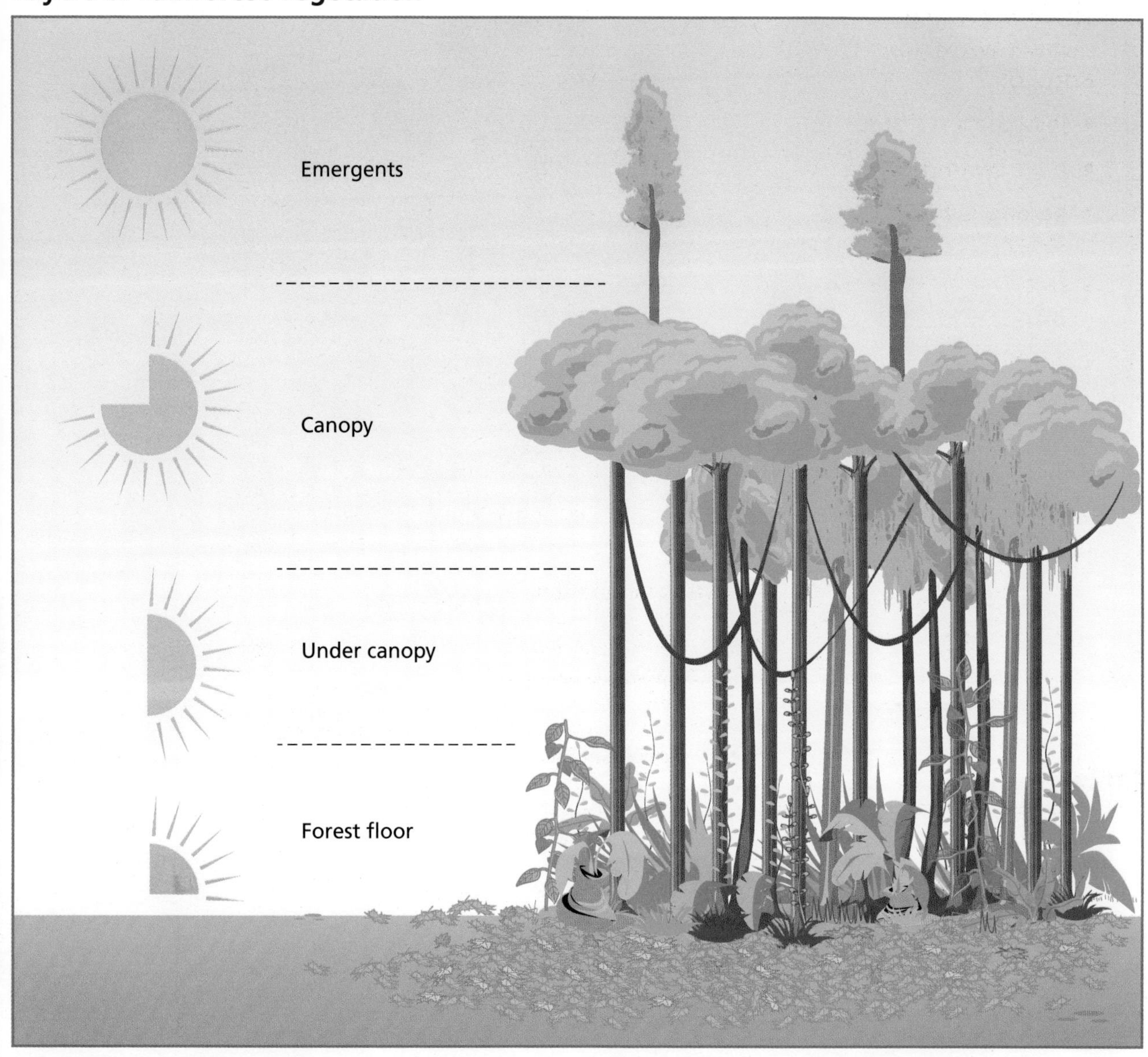

Rainforest water cycle

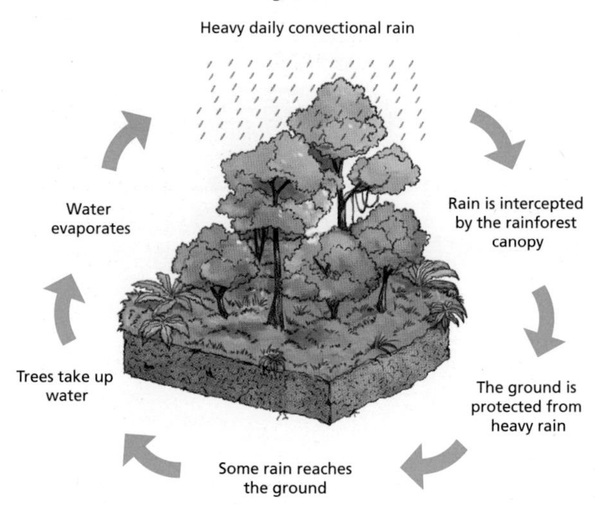

Rainforest nutrient cycle

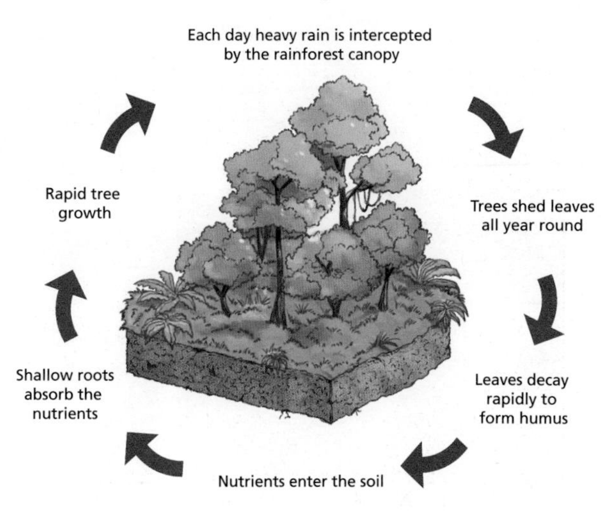

Forest cover on Borneo

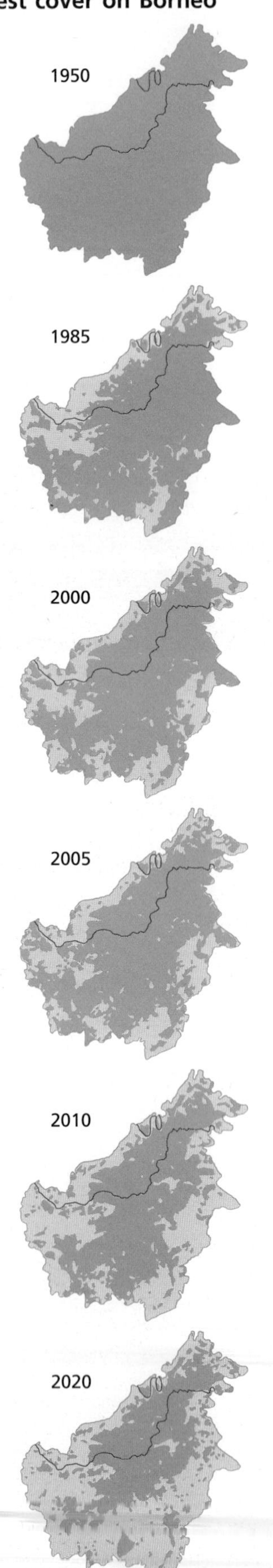

6.3 What is happening to tropical rainforests?

You should now be secure in your knowledge and understanding of why Borneo is a special place. Its stunning biodiversity contains many unique species and many more that have yet to be discovered. These are often very important resources for humans. Forests like those found on Borneo and in the Amazon also play an important role in regulating the earth's climate by capturing CO_2 as they grow, and storing it in plant tissue. All forests store more **carbon** than any other type of land cover and tropical forests can contain four times more carbon per hectare than cropland. These forests are vital in helping to absorb the millions of tonnes of CO_2 that humans create each year. They also provide benefits such as flood prevention, biodiversity and rainfall for farming.

However, these forests are under threat as they are being cut down, or deforested, at an increasing rate. An area of tropical forest thirteen million hectares in size is cut down and converted to other land uses each year. This is the equivalent of about 8.5 million football pitches a year, or 23,483 pitches a day. Figure A shows a landscape destroyed by logging in Sarawak, Borneo.

The reasons for deforestation are complex. In the past, forests were cut down to provide land to grow crops and to graze cattle for **subsistence farming**. Today, the land is used increasingly to provide products demanded on world markets, such as soy, **palm oil**, beef, timber, biofuels and minerals like gold, copper and **bauxite**. Deforestation occurs when it is cheaper to supply products from converted forest land than from other land. Our demand for products like palm oil and beef push the price higher and this leads to more forest being cut down to provide these products. The price of forest land doesn't yet include the value of forests in absorbing carbon from the atmosphere. Until this is added to the price of the land, they will continue to be cut down at increasing rates.

Deforestation is devastating for forest ecosystems. The habitats of animals like orang-utans are being destroyed and they are under threat of extinction. Undiscovered plants that might contain valuable medicines are being lost forever. Forest tribes like the **Penan** of Borneo in Malaysia are being broken up and are losing the land they have lived in harmony with for generations.

Figure A

Consolidating your thinking

Geographers use **Geographical Information Systems (GIS)** to monitor changes to places and there are networks of scientists around the globe using data from the ground and remote sensing data from satellites to track changes to forests. You will need to use your geographical skills to analyse the maps of Borneo forest loss from 1950 onwards on the opposite page. The following steps will help you to analyse a map like this:

- Describe the pattern of forest loss by using specific locations, e.g. 'Kalimantan', and directions, e.g. 'northeast'.
- Use the scale in km and time periods key (1950, 1950–2000, 2000–2020) and observe the projected forest cover in 2020.
- Describe which countries have the most and least forest.
- Describe what sort of forest has been lost by cross referencing with maps from the previous section.

Once you have done this, you are going to focus in on Sarawak and the changes from 2005 to 2010. You will need to download the Google Earth layer from http://earth.sarvision.nl/Sarawak_forest_change_2005-2010.kmz.

You should use Google Earth to switch these layers on and off to explore the relationship between forest loss, **plantations**, parks and reserves and the satellite photos underneath. You can also look at older satellite images to compare the change over time. There is a help sheet available for you in the Teacher Book (p.64). Take screenshots and paste these into a document with your annotations to build up a picture of what is happening in Sarawak. An annotated screen shot has been provided for you in the Teacher Book on page 65.

6.4 What is the invisible ingredient in deforestation?

Huge areas of valuable tropical rainforest are being cut down and burned, orang-utans are losing their habitats and are being pushed towards extinction and ancient peat bogs are being destroyed, releasing huge amounts of methane and CO_2 into the atmosphere.

Borneo's most famous mammal is the orang-utan, a great ape that shares 97% of its DNA with humans. In 1900, there were around 315,000 orang-utans. Today, fewer than 50,000 exist in the wild, split into small groups with little long-term chance of survival.

This destruction is driven by the global demand for palm oil – oil produced from the fruit of the oil palm tree that grows in tropical climates. Palm oil is found in an astonishing range of products. Here is a sample: margarine, cereals, crisps, sweets, baked goods, soaps, washing powders, fish fingers, pasties, pizzas, chewing gum and even cosmetics.

A palm oil plantation in Malaysia

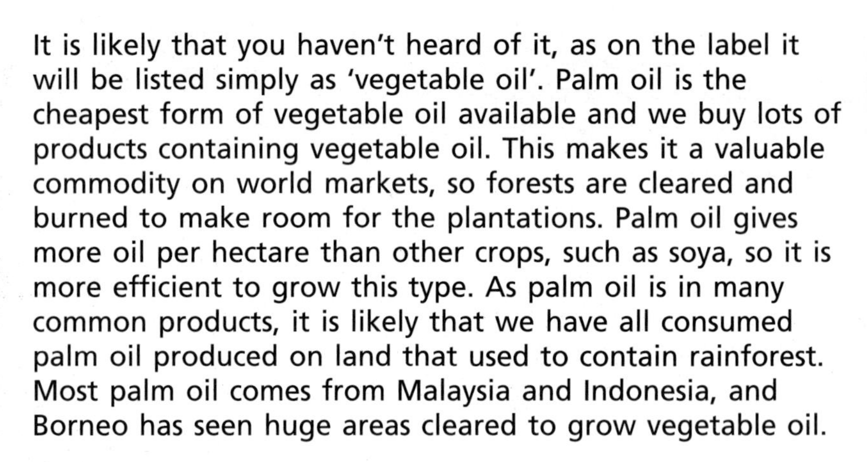

It is likely that you haven't heard of it, as on the label it will be listed simply as 'vegetable oil'. Palm oil is the cheapest form of vegetable oil available and we buy lots of products containing vegetable oil. This makes it a valuable commodity on world markets, so forests are cleared and burned to make room for the plantations. Palm oil gives more oil per hectare than other crops, such as soya, so it is more efficient to grow this type. As palm oil is in many common products, it is likely that we have all consumed palm oil produced on land that used to contain rainforest. Most palm oil comes from Malaysia and Indonesia, and Borneo has seen huge areas cleared to grow vegetable oil.

Unless the deforestation is halted, all of the rainforest will ultimately disappear. Consumers like you have the power to bring about change by demanding products that contain sustainable palm oil. This product is widely available and ensures that strict global sustainability standards are met. In 2013, Wilmar International – the world's largest palm oil company – committed to a policy of zero deforestation. Other major buyers are now making pledges to use only sustainably sourced oil, but there is a considerable distance to go to completely halt deforestation for this purpose and to avoid the environmental madness of destroying tropical forests to make margarine!

Palm oil facts:

- It is found in thirty-two of the sixty-two biggest-selling food brands in the UK.
- Fifty million tonnes are produced globally each year.
- Around 85% of the world's palm oil is from Malaysia and Indonesia.

From forest to factory

1. *Newly cleared and planted forest*

2. *A major plantation*

3. *A labourer carries the fruit from the palm tree*

4. *A farmer takes his crop to sell*

5. *Unloading the heavy fruits by hand*

6. *Mr Willi has only recently started his purchasing business*

7. *One day's worth of deliveries*

8. *The weekly price varies due to global prices*

9. *The individual fruits that are squeezed and heated*

10. *A lorry load is sent to the processing plant*

11. *Inside the processing plant*

12. *The large scale processing plant on cleared forest*

Consolidating your thinking

Both the people and the governments of Malaysia and Indonesia want their economies to grow, employment prospects to increase and quality of life to improve. Plantations have existed on Borneo since they were established by the British toward the end of the nineteenth century, so not all of the palm oil is being grown on newly deforested areas.

Are the modern palm oil plantations sustainable? You will need to consider this question from a range of perspectives. Think about the farmer in the photo sequence – palm oil provides him with an income and helps him feed and educate his children. His palm fruits are grown on land deforested nearly 100 years ago. Is this a sustainable form of income for him and his family?

What about the middle man, Mr Willi? He takes palm oil from all the local small-scale farmers – regardless of where it is grown – and sells it on to the local processing plant. He wants to make money, but his prices are controlled by the global demand. He is one of hundreds of individuals that have set themselves up in Sarawak near Lambir Hills. Are *they* driving the small-scale destruction of forests by individual farmers or are *we*, by continuing to purchase products with unsustainable palm oil?

Consider some of the economics involved. The price for 1000 kg of grade A oil palm fruit (the largest type, grown for the longest time) is 360 Malaysian Ringgit (MYR) – approximately US$104. It takes thirty months for the trees to start bearing fruit. A typical hectare could contain 100 trees and these could produce approximately four tonnes of crude palm oil per year. This equates to US$415 per hectare of oil palm per year.

You will need to use your own research and the information here to produce the text and images for a breakfast cereal company that uses sustainable palm oil in their new product. They want to provide consumers with information about both palm oil-led deforestation, its impacts on biodiversity and how sustainable practices are making a difference. You should use digital technology to assist you with the design, remembering that this is about showing your geographical understanding of the issue, the concept of sustainability and the strength of consumer power.

Web links to help with your research:

http://news.nationalgeographic.com/news/2014/10/141009-orangutans-palm-oil-malaysia-indonesia-tigers-rhinos/

http://www.rspo.org/consumers/about-sustainable-palm-oil

http://www.greenpeace.org.uk/blog/forests/palm-oil-companies-say-theyll-put-forest-destruction-hold-what-happens-next-20140919

http://www.theguardian.com/environment/2014/jun/29/rate-of-deforestation-in-indonesia-overtakes-brazil-says-study

http://www.greenpalm.org/

Extending your enquiry

There are many **indigenous people** in remote areas of Borneo and their traditional way of life has been significantly altered by deforestation. One tribe, the Penan, were traditionally nomadic hunter-gatherers that moved around the forest, only taking what they needed before moving on. Over the last thirty years they have become more fixed in location and often their only form of income is to work for logging companies or palm oil plantations. Many of the Penan now live in poverty as they are excluded from the profits of the palm oil industry. Their levels of education are low and they often lack access to basic healthcare.

Is economic development on Borneo sustainable for the Penan?

The images here show a group of teachers that work on Borneo and are working with the Penan to sell their handmade baskets and goods. These beautiful products are made by the Penan women out of rattan from the forest or from recycled plastic. The teachers have been raising funds for years and provide the Penan children with school fees, uniforms, equipment and safe transport. One Penan girl is now at university studying Public Administration Science, supported by the students of Jerudong International School. Lonnie is the first Penan from the Limbang District ever to enter tertiary education.

The Penan are at the forefront of the need for sustainable economic development on Borneo. You should research their tribe, the area they inhabit and how they traditionally used the forest. You should also find out about the threats they face and how this can be managed.

You could also research the work of conservation organisations and the attempts that are being made to protect and restore the forests on Borneo. Some of these are global initiatives and others are more local:

- The WWF Heart of Borneo Project
- The Sepilok Orangutan Rehabilitiaion Centre
- OuTrop – Orangutan Tropical Peatland Project
- The Kinabatangan Corridor of Life

For the Penan, or for one of the examples above, you could produce your own short enquiry, starting with a mind map of questions, then selecting, prioritising, researching and producing your work in a format of your choice – video, presentation, poster or written report.

Web links to help with your research:

http://www.survivalinternational.org/tribes/penan

http://www.bbc.co.uk/programmes/p009r19p

http://www.bbc.co.uk/tribe/tribes/penan/index.shtml

Glossary

A

abstraction the removal of water from any source. This water can be used for irrigation, industry, recreation, flood control or treatment to produce drinking water

adaptation the process by which living organisms, including humans, change the way they live to survive better in an environment

aid (international) money, goods and services given by the government of one country or a multilateral institution, such as the World Bank or International Monetary Fund (IMF), to help another country

aquifer an underground layer of permeable rock that contains water. The water can be abstracted using boreholes and wells

arid having little or no rain; too dry or barren to support vegetation

arid subtropical climate climate zones characterised by an annual average temperature of 18.2 °C (64.8 °F), a lack of regular rainfall and high humidity. These are generally located in areas next to powerful cold ocean currents, e.g. the coast of Peru and the coastal areas of southern Africa (Namibia, South Africa)

aridity index a numerical indicator of the degree of dryness of the climate at a given location

atmosphere the air which surrounds the earth consisting of three layers: the troposphere, mesosphere and ionosphere

B

barriada a shantytown or slum section on the outskirts of a large city in Latin America. Lima in Peru has large Barriadas

bauxite an ore used in the manufacture of aluminium

beach combers the recreational activity of looking for and finding various interesting objects that have washed in with the tide: seashells, fossils, pottery shards, historical artefacts and driftwood. Items lost from some seagoing vessels can also be collected and can lead to scientific discoveries

biodegrade the process where a waste product decays and becomes absorbed by the environment

biodiversity the variety of plant and animal life in the world or in a particular habitat

biosphere the part of the earth which contains living organisms. The biosphere contains a variety of habitats, from the highest mountains to the deepest oceans

borehole a deep hole drilled down into the ground when looking for oil, gas or water

Borneo the third-largest island in the world and the largest island of Asia. It comprises three countries: Brunei, Indonesia (Kalimantan) and Malaysia (Sabah and Sarawak)

brain drain the emigration of scientists, technologists, academics, etc. for better pay, equipment, or conditions

bycatch the unwanted fish and other marine creatures trapped by commercial fishing nets during fishing for a different species

C

carbon storage the absorption of carbon dioxide (CO_2) by plants as they grow. Plants store CO_2 and release oxygen through photosynthesis

cartographer a person who creates maps

climate change changes in the earth's weather, especially the increase in the temperature of the earth's atmosphere that is caused by the increase of particular gases, especially carbon dioxide

coastal plain an area of flat, low-lying land next to the coast

colonise to settle in an area and take control

composite in maps this is to combine two or more maps to make a single map

condensation the process by which cooling vapour turns into a liquid. Clouds, for example, are formed by the condensation of water vapour in the atmosphere

conservation the careful use and management of natural resources to ensure their sustainable exploitation and prevent them from being lost or wasted

constraint something which sets a limitation or restriction

consumption an umbrella term for the many different ways and rates that humans consume the products of the natural world

container ship cargo ships that carry all of their load in truck-size containers, in a technique called containerisation. Most seagoing cargo of goods is carried on these ships. The 400 m long Mærsk Triple E class ships can hold 18,340 twenty-foot containers

coral reefs diverse underwater ecosystems held together by calcium carbonate structures secreted by corals. The Great Barrier Reef in Australia is a famous example

correlation in statistics this is a relationship or

connection between two or more variables (things) which can be positive (as one thing increases so does the other) or negative (as one thing increases the other decreases)

cultivation to prepare and work on land in order to grow crops

cyclone a wind which has a velocity of more than 118 km per hour. Cyclones can cause great damage by wind as well as from the storm waves and floods that accompany them

D

deforestation the practice of clearing trees. Much deforestation is a result of development pressures, e.g. trees are cut down to provide land for agriculture and industry

desertification the encroachment of desert conditions into areas which were once productive. Desertification can be due partly to climatic change, i.e. a move towards a drier climate in some parts of the world (possibly due to global warming), though human activity has also played a part through bad farming practices. The problem is particularly acute along the southern margins of the Sahara Desert in the Sahel region between Mali and Mauritania in the west, and Ethiopia and Somalia in the east

drought a long period of unusually low rainfall that can lead to a shortage of water

E

economic development the building up of prosperity and wealth of countries, regions or communities for the benefit of the people who live there

economic water scarcity a type of water scarcity caused by a lack of investment in water or where the population does not have the necessary monetary means to access the water that is available

ecoregion an area defined by its environmental conditions, especially climate, landforms, soil characteristics and vegetation

equatorial climate a climate in which there is no dry season and no obvious summer or winter – it is typically hot and wet throughout the year and rainfall is both heavy and frequent – and all months have an average precipitation value of at least 60 mm. One day in an equatorial climate can be very similar to the next, while the change in temperature between day and night may be larger than the average change in temperature along the year

erosion the wearing away of the earth's surface by running water (rivers and streams), moving ice (glaciers), the sea and the wind

Eurozone crisis a shorthand term which refers to when six countries in Europe had to be saved from bankruptcy by other countries, who contributed money so that they could pay their debts

evapotranspiration the return of water vapour to the atmosphere by evaporation from land and water surfaces and by the transpiration of vegetation

export part of international trade where goods produced in one country are shipped to another country for future sale or trade

F

fertiliser any substance, such as manure or a mixture of nitrates, added to soil or water to increase its productivity

flash flood a sudden and destructive rush of water down a narrow gully or over a sloping surface, caused by heavy rainfall

fish factory a large fishing vessel equipped with processing and freezing equipment. The largest of these have forklift trucks that can off load the fish to refrigeration vessels for return to shore

flood an overflow of a large amount of water beyond its normal limits, especially over what is normally dry land

flotsam debris in the water that was not deliberately thrown overboard, often as a result of a shipwreck or accident. The word flotsam derives from the French word floter, to float. Under maritime law flotsam may be claimed by the original owner

fog catcher very large screens constructed in arid areas. As fog drifts in, water droplets condense around the thin screens and drip to the collection pools below. In one day, a single screen can collect more than a hundred gallons of water

G

G8 Summit the G8 are a group of eight of the world's most powerful countries: Canada, France, Germany, Italy, Japan, Russia, the UK and the USA. The leaders of these countries meet once per year at the G8 Summit

garbage patch an area of marine debris concentration in the Pacific and Atlantic oceans

Geographical Information System (GIS) digital base maps with different layers of information added over the top. They are used by geographers to display information and patterns, identify issues and make decisions. Google Earth is a basic example of GIS

glacier a body of ice occupying a valley and originating in a corrie or icefield. A glacier moves at a rate of several metres per day, the precise speed depending upon climatic and topographic conditions in the area in question

global warming the warming of the earth's atmosphere caused by an excess of carbon dioxide, which acts like a blanket, preventing the natural escape of heat. This situation has been developing over the last 150 years because of (a) the burning of fossil fuels, which releases vast amounts of carbon dioxide into the atmosphere, and (b) deforestation, which results in fewer trees being available to take up carbon dioxide

global water resources the total amount of water available on Earth. This is held and transferred between different stores and is not evenly distributed. 97% of all water is in the seas and oceans

Greenpeace an independent global campaigning organisation that acts to change attitudes and behaviour, to protect and conserve the environment and to promote peace

gross domestic product (GDP) the total value of all goods and services produced domestically by a nation during a year

gyres a large system of rotating ocean currents associated with large wind movements and the earth's rotation, e.g. the North Atlantic Gyre

H

high-tech an abbreviation for high-technology which generally refers to products incorporating the most advanced forms of computer electronics

human development index (HDI) a statistical tool used to measure a country's overall social and economic achievement

hydroelectric power the generation of electricity by turbines driven by flowing water. Hydroelectricity is most efficiently generated in rugged topography where a head of water can most easily be created, or on a large river where a dam can create similar conditions. Whatever the location, the principle remains the same – that water descending via conduits from an upper storage area passes through turbines and thus creates electricity

I

ice cap a covering of permanent ice over a relatively small land mass, e.g. Iceland

indigenous something produced, growing, living, or occurring naturally in a particular region or environment

infiltration the process by which water on the ground surface enters the soil

indigenous people people who have specific rights based on their historical and cultural ties to a particular place or territory

interconnection how places are linked

interdependent where people and countries are dependent on one another in some way. For example, in high income countries, people are dependent and linked to people in low income countries when they buy agricultural products that were have grown for export, like coffee or cocoa beans used in chocolate

irrigation a system of artificial watering of the land in order to grow crops. Irrigation is particularly important in areas of low or unreliable rainfall

J

jetsam debris that was deliberately thrown overboard by a crew of a ship in distress, most often to lighten the ship's load. Jetsam is a shortened word for jettison. Under maritime law jetsam may be claimed as property of whoever discovers it

M

malnutrition a lack of proper nutrition, caused by not having enough to eat, not eating enough of the right things or being unable to use the food that you do eat

marine litter (marine debris) human-created waste that has deliberately or accidentally been released in a lake, sea, ocean or waterway. Floating oceanic debris tends to accumulate at the centre of gyres and on coastlines, frequently washing ashore

migration the movement of people or animals involving a permanent or semi-permanent change of residence

multidimensional poverty index (MPI) a measure which uses different factors to determine poverty beyond income-based lists. It allows for broader comparisons both across countries, regions and the world and within countries by ethnicity, urban/rural location and other factors.

multiplier effect when an investment of money

in one thing or project leads to an increase in jobs, income and consumption much greater than that created by the original amount spent

N

NASA stands for National Aeronautics and Space Administration and is the US government agency responsible for science and technology related to air and space

natural disaster a natural event which, in extreme cases, can lead to loss of life and destruction of property. Some natural disasters result from geological events, such as earthquakes and the eruption of volcanoes, whilst others are due to weather events, such as cyclones, floods and droughts

natural services the services provided by ecosystems, such as the provision of clean drinking water, decomposition of waste, crop pollination, climate regulation and carbon storage

natural systems interlinked systems that make up the natural environment, e.g. the climate system, the biosphere system, the hydrological system and the lithosphere system (soil, rocks and the crust)

nomadic a member of a community of people who live in different locations, moving from one place to another; often associated with livestock migration or hunting and gathering

nutrient pollution the process where too many nutrients, mainly nitrogen and phosphorus, are added to water and can act like fertiliser, causing excessive growth of algae. Nutrients can run off from agricultural land into rivers

O

ocean current a movement of the surface water of an ocean

oceanographer a scientist that studies the oceans, the life that inhabits them and their physical characteristics, e.g. depth and currents

Olmos Irrigation Project a large scale irrigation project in Northern Peru that involves the transfer of water across the Andes through a network of dams and tunnels. The water will create 38,000 hectares of newly irrigated farmland at an estimated cost of US$500 million

OSCURS the Ocean Surface Current Simulator. A computer model that measures the movement of ocean surface currents over time

Ottoman Empire a large Turkish empire that existed between the fourteenth and twentieth centuries. Its capital was Constantinople (now Istanbul)

overfishing to deplete the stock of fish in an area of sea by fishing excessively

P

palm oil an edible vegetable oil derived from the fruit of the oil palms tree, primarily the African oil palm. Indonesia and Malaysia are the largest exporters of palm oil in the world. Palm oil is used in hundreds of every day products we use and consume. Areas of tropical rainforest are being cleared for new plantations

pastoralism the branch of agriculture concerned with the raising of livestock

pelagic trawler a boat that tows giant cone-shaped nets in the middle of the water column to catch fish such as tuna, mackerel, anchovies and shrimp

Penan a nomadic aboriginal people living in Sarawak and Brunei, on Borneo. The Penan number around 16,000 – of which only approximately 200 still live a nomadic lifestyle. The Penan are noted for their practice of 'molong' which means never taking more than necessary

percolate percolation is where water moves downward from the soil into porous rock

pesticide normally a chemical substance used to destroy pests such as insects (insecticides) and weeds (herbicides)

photodegrade where an object is decomposed by the action of light, especially sunlight

physical water scarcity the situation where there is not enough water to meet all demands, including that needed for ecosystems to function effectively

pirate fishing illegal, unreported and unregulated fishing that takes place in violation of rules (e.g. the number of fish that can be caught) or operate in waters without permission

plantation an estate on which crops such as rubber, coffee, sugar or palm oil are grown

pluvial flooding occurs when saturated land cannot absorb further rainfall, which then flows out over the surrounding rural or urban land

population density the average number of people living in each square kilometre of an area

porous rocks rocks which are more likely to absorb water because of their rounded grains. Porous rocks tend to be softer and more crumbly than non-porous rocks with interlocking grains.

precipitation water deposited on the earth's surface in the form of, e.g. rain, snow, sleet, hail and dew

R

region an area that has definable characteristics which make it distinct

S

sanitation the provision of clean drinking water and sewage/waste water disposal and cleaning systems

Silicon Valley the nickname of the southern part of San Francisco Bay in California, USA where many of the world's largest high-tech businesses and corporations can be found

Sub-Saharan Africa geographically it is the area of the continent Africa that lies south of the Sahara Desert. Politically it consists of all African countries that are located south of the Sahara

subsistence farming a system of agriculture in which farmers produce exclusively for their own consumption, in contrast to commercial agriculture where farmers produce purely for sale at the market

sustainability development that meets the needs of the present without impacting negatively on the environment or compromising the ability of future generations to meet their own needs

T

topography the natural features which make up the surface of the land

tropical rainforest the dense forest cover of the equatorial regions, reaching its greatest extent in the Amazon Basin of South America, the Congo Basin of Africa, and in parts of Southeast Asia and Indonesia. There has been much concern in recent years about the rate at which the world's rainforests are being cut down and burned. The burning of large tracts of rainforest is thought to be contributing to global warming. Many governments and conservation bodies are now examining ways of protecting the remaining rainforests, which are unique ecosystems containing millions of plant and animal species

U

United Nations an intergovernmental organisation which was founded in 1945. The main aim of the UN was to prevent another conflict like World War II. There are 193 UN member states and the current Secretary-General is Ban Ki-moon

urban a built up area such as a town or city

V

virtual water the sum of all the water used in the various steps of the production chain of a given product, e.g. to produce 1kg of wheat we need about 1000 litres of water

W

water consumption the amount of water consumed by an individual in one day. Usually measured in litres

waterhole a depression in the ground containing water, such as a pond or pool, used by animals as a drinking place

water-stress occurs when the demand for water exceeds the available amount during a certain period or when poor quality restricts its use. Water-stress causes a reduction of fresh water quantity and quality and is a major constraint on human activity

water table the level below which the ground is saturated with water

water transfer scheme where water is moved from one drainage basin to another for human consumption, agriculture or industry

WWF the World Wide Fund for Nature is working on issues regarding the conservation, research and restoration of the environment

Index

Note: page numbers in bold refer to maps.

Index

Acknowledgements

The publishers wish to thank the following for permission to reproduce photographs, illustrations and other graphics. Every effort has been made to trace copyright holders and to obtain their permission for the use of copyright materials. The publishers will gladly receive any information enabling them to rectify any error or omission at the first opportunity.

Illustration by Jouve Pvt Ltd p.82

Images: (t = top, c= center, b= bottom, r = right, l = left)
Cover and tile page image © Pichugin Dmitry/Shutterstock.com; p4 (tl) © DLR German Aerospace Centre/Flickr.com CC BY 3.0; p4 (tc) © Johnson Space Center/NASA Public domain; p4 (c) © ESA/Rosetta/NAVCAM/Wikimedia Commons CC BY-SA IGO 3.0; p4 (bc) © NASA/Kennedy Space Center/; p5 (t) © NASA Public domain; © DLR German Aerospace Centre/Flickr.com CC BY 3.0; p5 (r) © Zoetnet/Flickr.com CC BY-SA 2.0; p8 © Muhammad Mahdi Karim/Wikimedia Commons GNU Version 1.2; p9 (t) © Mudjo/Wikimedia Commons CC BY-SA 3.0; p9 (b) © idobi/Wikimedia Commons CC BY-SA 3.0; p10 (t) ©Angrense/Wikimedia Commons Public domain; p10 (c) © Olga Popova/Shutterstock.com; p10 (b) © Joseph August/Shutterstock.com; p11 ©Melchisedech Thevenots, *Relations de divers voyages curieux*, Paris, J. Public domain; p12 © Laborant/Shutterstock.com; p14 © Arthur Hidden/Shutterstock.com; p15 (tl) © Konstantin Stepanenko/Shutterstock.com; p15 (tr) © Zhukov Oleg/Shutterstock.com; p15 (b) © Hitdelight/Shutterstock.com; p20 (t) © Philip Lange/Shutterstock.com; p20 (b) © vitamind/Shutterstock.com; p21 © ChameleonsEye/Shutterstock.com; p22 © Harvepino/Shutterstock.com; p24 (tl) © Tracey Williams; p24 (b) © LegoLostAtSea/Wikimedia Commons Public Domain; p25 (tr, cl, cr, cbr, bl) © Tracey Williams; p25 (br) © New Zealand Defece Force/Wikimedia Commons CC BY 2.0; p26 (t) © Commission Air/Alamy; p26 (cl, cr, bl, br) © Tracey Williams; p28 © Rick Rickman/NOAA Public domain; p30 © Tracey Williams; p31 (t) © Marine Debris/NOAA Public domain; p31 (cr) © Rosanne Tackaberry/Alamy; p32 (tl) © Ethan Daniels/Shutterstock.com; p32 (tr) © Tracey Williams; p32 (cl, cr, b) © Ambient Images Inc./Alamy; p34 © Ambient Images Inc./Alamy; p35 (tr) © C. Ortiz Rojas/NOAA Public domain; p35 (b) © Greenpeace/Paul Hilton; p36 © Greenpeace; p38 (tl) © Nesnad/Wikimedia Commons CC BY-SA 4.0; p38 (cl) © World-spec/NASA/Alamy; p39 (cl) © Kennedy Space Center/NASA Public domain; p39 (r) © Johnxxx9/Wikimedia Commons CC BY-SA 3.0; p39 (bl) © Nesnad/Wikimedia Commons Public domain; p39 (br) © epa european pressphoto agency b.v./Alamy; p41 (tr) © Antônio Milena/Abr/Wikimedia Commons CC BY-SA 3.0; p41 (tcr) © Dennis Jarvis/Wikimedia Commons CC BY-SA 2.0; p41 (cr) © meg and rahul/Wikimedia Commons CC BY-SA 2.0; p41(cr) © Manuspanicker/Wikimedia Commons Public domain; p41 (bcr) © Abhishek727/Wikimedia Commons CC BY-SA 3.0; p41 (br) © JeremyRichards / Shutterstock.com; p44 (tl) © Bastian Greshake/Flickr.com CC BY-SA 2.0; p44 (cl) © Gregory H. Revera/Wikimedia Commons CC BY-SA 3.0; p44 (cr) © Jet Propulsion Laboratory/NASA Public domain; p44 (bl) © Sivavkm/Wikimedia Commons CC BY-SA 3.0; p45 © Roketjack/Wikimedia Commons Public domain; p46 (tl) © Mohseen Khan/Wikimedia Commons CC BY-SA 3.0; p46 (cl) © Razimantv/Wikimedia Commons CC BY-SA 3.0; p46 (bl) © EugenP/Shutterstock.com; p47 (tl) © Kalyan Shah/Wikimedia Commons CC BY-SA 3.0; p47 (tr) © Prateek Karandikar/Wikimedia Commons CC BY-SA 4.0; p47 (bl) © Thomas Imo/Alamy; p47 (cr) © Danita Delimont/Alamy; p47 (br) © Dinodia Photos/Alamy; p48 (tl) © Mike Goldwater/Alamy; p48 (tcl) © Dinodia Photos/Alamy; p48 (cr) © Supportstorm/Wikimedia Commons Public domain; p48 (bcl) © Mike Goldwater/Alamy; p48 (bl) © Stocktrek Images, Inc./Alamy; p48 (br) © NOAA Public domain; p49 (cl) © Amit/Wikimedia Commons CC BY-SA 2.0; p49 (cr) © Andreina Lairet/Flickr CC BY-SA 2.0; p49 (br) © Tech. Sgt. Keith Brown/U.S. Air force/Wikimedia Commons Public domain; p50 (t) © Baciu/Shutterstock.com; p50 (b) © Tukaram Karve/Shutterstock.com; p51 (tr) © Kennedy Space Center/NASA Public domain; p51 (cl) © Johnson Space Center/NASA Public domain; p51 (cr) © MrMiscellanious/Wikimedia Commons Public domain; p51 (cr) © Johnson Space Center/NASA Public domain; p51 (bl) © Spinoff.nasa.gov/NASA Public domain; p51 (br) © Johnson Space Center/NASA Public domain; p52 (tl) © auremar/Shutterstock.com; p52 (tr) © Panos Karas/Shutterstock.com; p52 (cl) © John Karakatsanis/Flickr CC BY-SA 2.0; p52 (cr) © A.P.S. (UK)/Alamy; p52 (bl) © Steve Garvie/Wikimedia Commons CC BY-SA 2.0; p52 (br) © Eric Inafuku/Wikimedia Commons CC BY-SA 2.0; p53 (tl) © Natursports/Shutterstock.com; p53 (tr) © ronfromyork/Shutterstock.com; p53 (cl) © Combined Joint Task Force - Horn of Africa/U.S. Air Force Public domain; p53 (cr) © pavla/Shutterstock.com; p53 (bl) © NASA Earth Observatory Public domain; p53 (br) © SandiMako/Shutterstock.com; p55 (t) © iofoto/Shutterstock.com; p55 (b) © Ambrophoto/Shutterstock.com; p56 (tl) © Sam DCruz/Shutterstock.com; p56 (cl) © Sam DCruz/Shutterstock.com; p56 (bl) © Sean Sprague/Alamy; p57 (tr, cr, cl, br) © Sean Sprague/Alamy; p58 (tl, cl, bl) © Sean Sprague/Alamy; p58 (br) © Geophysical Fluid Dynamics Laboratory/NOAA Public domain; p59 (t) © Images of Africa Photobank/Alamy; p59 (tr) © Oscar Max/Alamy; p59 (c) © Sue Cunningham Photographic/Alamy; p59 (br) © Mark Boulton/Alamy; p59 (b) © a katz/Shutterstock.com; p60 (t) © Peretz Partensky/Wikimedia Commons CC BY-SA 2.0; p60 (b) © Mwtoews/Wikimedia Commons GNU Version 1.2; p61 (t) © Isewell/Wikimedia Commons CC BY-SA 2.0; p61 (b) © Wikimedia Commons Public domain; p62 (t) © Bureau of Reclamation/U.S. Federal Government Public domain; p62 (cl) © Ronnie Macdonald/Wikimedia Commons CC BY 2.0; p62 (cr) © U.S. Federal Government Public domain; p62 (b) © Rob Young/Wikimedia Commons CC BY 2.0; p64 (bl, br) © Photoshot Holdings Ltd/Alamy; p65 (tr, tr) © Photoshot Holdings Ltd/Alamy; p65 (cl) © John Warburton-Lee Photography/Alamy; p65 (cr) © National Geographic Image Collection/Alamy; p65 (br) © Images of Africa Photobank/Alamy; p66 (t) © USDA Public domain; p66 (c) © Africa Media Online/Alamy; p66 (b) © Alex Hinds/Alamy; p67 (tr) © Staecker/Wikimedia Commons Public domain; p67 (c) © NASA Public domain; p67 (br) © Jeff Schmaltz, MODIS Land Rapid Response Team/NASA Public domain; p69 (t, tr, c, br, b) © Dartmoor National Park Authority; p71 © Dartmoor National Park Authority; p72 (t) © Yongyut Kumsri/Shutterstock.com; p72 (b) © Martchan/Shutterstock.com; p73 (b) © Andrea Izzotti/Shutterstock.com; p77 (t) © Scorpp/Shutterstock.com; p77 (tr) © Mariemily Photos/Shutterstock.com; p77 (tcr) © Aggie 11/Shutterstock.com; p77 (bcr) © Pablo Hidalgo - Fotos593/Shutterstock.com; p77 (br) © Aleph Studio/Shutterstock.com; p77 (b) © ProKasia/Shutterstock.com; p78 (tl, tr, tcl, bcl, bl) © Jake Lyell/Alamy; p79 © Universal Images Group Limited/Alamy; p80 © 2014 Inav/Geosistemas SRL/Google Earth Public domain; p81 (t) © Indisdepe/Wikimedia Commons CC BY-SA 3.0; p81 (b) © USGS Public domain; p82 (t) © 2014 DigitalGlobe/Google Earth Public domain; p82 (bl) © Image Source/Alamy; p83 © Anne Lummerich; p84 (tl) © Designua/Shutterstock.com; p84 (tr) © Kaleidoscopio/Shutterstock.com; p84 (bl) © Stephen Chung/Shutterstock.com; p86 (tl) © Christian Vinces/Shutterstock.com; p86 (tr) © Tom Grundy/Shutterstock.com; p86 (cl) © Tim Roberts Photography/Shutterstock.com; p86 (bl, br) © 2014 DigitalGlobe/Google Earth; p91 (t) © UbjsP/Shutterstock.com; p91 (b) © Kim Briers/Shutterstock.com; p92 (b) © Fabio Lamanna/Shutterstock.com; p94 © MODIS Rpid Response System/NASA Public domain; p96 (t) © Nicholas Sheehan; p96 (cl) © sydeen/Shutterstock.com; p96 (cr) © Fabio Lamanna/Shutterstock.com; p96 (bl) © Ryan M. Bolton/Shutterstock.com; p96 (bc) © Michal Ninger/Shutterstock.com; p96 (br) © Nicholas Sheehan; p97 (tl) © WayneImage/Shutterstock.com; p97 (tr) © Nicholas Sheehan; p97 (cl) © Kim Briers/Shutterstock.com; p97 (cr) © Praisaeng/Shutterstock.com; p97 (bl) © szefei/Shutterstock.com; p97 (br) © Nokuro/Shutterstock.com; p99 (tl) © Dr. Morley Read/Shutterstock.com; p99 (tr) © A.S. Zain/Shutterstock.com; p99 (cr) © Rich Carey/Shutterstock.com; p99 (br) © Mint Images Limited/Alamy; p100 © Nature Picture Library/Alamy; p101 (tl, tr, tcl, tcc, tcr, bcl, bcc, bcr, bl) © David Weatherly; p101 (tr) © wandee007/Shutterstock.com; p101 (bc) © KYTan/Shutterstock.com; p101 (br) © Meister Photos/Shutterstock.com; p103 (tr, cr, br) © Dom Powles/Jacky McLaren